A HISTORY OF THE
HOPE ENTOMOLOGICAL COLLECTIONS
IN THE
UNIVERSITY MUSEUM OXFORD

Oxford University Museum
Publication 2

A HISTORY OF THE
Hope Entomological Collections
IN THE
University Museum Oxford

WITH LISTS OF ARCHIVES
AND COLLECTIONS

BY

AUDREY Z. SMITH
Hope Librarian

CLARENDON PRESS · OXFORD
1986

Oxford University Press, Walton Street, Oxford OX2 6DP
Oxford New York Toronto
Delhi Bombay Calcutta Madras Karachi
Petaling Jaya Singapore Hong Kong Tokyo
Nairobi Dar es Salaam Cape Town
Melbourne Auckland
and associated companies in
Beirut Berlin Ibadan Nicosia

Oxford is a trade mark of Oxford University Press

Published in the United States
by Oxford University Press, New York

British Library Cataloguing in Publication Data

University Museum (Oxford)
A history of the Hope entomological collections
in the University Museum Oxford: with a list of
collections and archives.—(Oxford University
Museum publication ; 2)
1. University Museum (Oxford)—History
2. Hope, Frederick William—Private collections
3. Insects—Catalogs and collections
I. Title   II. Smith, Audrey Z.   III. Series
595.7'074'02574   QL468.2
ISBN 0-19-858179-3

Library of Congress Cataloging in Publication Data
Smith, Audrey Z.
A history of the Hope Entomological Collections in the
University Museum, Oxford.
(Publication/Oxford University Museum ; 2)
Bibliography: p.
Includes index.
1. Insects—Catalogs and collections—England—
Oxfordshire—History. 2. University of Oxford.
University Museum—Natural history collections—History.
I. Title.   II. Title: Hope Entomological Collections.
III. Series: Publication (University of Oxford.
University Museum) ; 2.
QL468.2.S65 1986   595.7'074'02574   85-28354
ISBN 0-19-858179-3

Printed in Great Britain
at the University Printing House, Oxford
by David Stanford
Printer to the University

Preface

THE desirability of providing a historical account of the Hope Entomological Collections, formerly the Hope Department of Zoology (Entomology), arose in 1947 in connection with a proposal to mark the centenary of the founding of the Entomological Collections. Unfortunately this did not come about. However, a talk given by Ernest Taylor, at the 248th meeting of the Oxford University Entomological Society held at Jesus College on 22 November 1963, led to a revival of the idea. The talk, based on notes compiled by Ernest Taylor, then Chief Technician, and myself, proved so popular that many requests were made for it to be published.

Over the last five years, with considerable encouragement from Dr M. W. R. de V. Graham (until 1981 Curator of the Entomological Collections), I have added extensively to those notes and it was decided to provide as complete an account of the Collections as possible. The historical part contains much information which lies in scattered papers or unpublished manuscripts and is not easy to obtain. The account of the unique Library and list of archives contains much hitherto unpublished information. These parts, together with the list of collections and donors, should prove an indispensable source of information to scholars and research workers both in the University and in the scientific world in general.

The Hope Department of Zoology was established with the endowment of a professorial chair in 1860. Later, in order to avoid any confusion with the Department of Zoology and Comparative Anatomy, Convocation modified the title 'Hope Department of Zoology' in January 1934 by adding the word 'Entomology' to indicate the main work of the department. The title 'Hope Department of Zoology (Entomology)' remained in use until October 1978 when the teaching and research side of entomology was integrated by decree with the Department of Zoology. The Collections and Library remained in the University Museum and then, for practical purposes, came under the care of the Committee for the Scientific Collections with the title 'Hope Entomological Collections'. This is a far cry from the mid nineteenth century when in that delightful book *The adventures of Mr. Verdant Green*[†] the Hope Department was referred to as 'The Bug and Butterfly'—a footnote explaining that this was the name given to Mr Hope's Entomological Museum recently presented to the University, and housed at the Taylor Institution.

Proceeds from the sale of this book are being used to support further Oxford University Museum Publications.

Oxford   A.Z.S.
1984

[†] Cuthbert Bede (Edward Bradley), *The adventures of Mr. Verdant Green, an Oxford freshman* (London, 1853), p. 259.

Acknowledgements

My sincere thanks to the following:

To Dr Graham for his unfailing help so freely given in the preparation of the manuscript, and for his loyalty and support over the past thirty years; his great knowledge of the Collections and the history of entomology has been absolutely invaluable.

To Mrs Ruth Wickett who has so ably assisted me in the Library for the past ten years, and for her help in so many ways in preparing the manuscript for publication.

To Derek Whiteley (artist in the Hope Department 1960–8) for his gift of photographic work and cover illustration.

To Professor D. Spencer Smith (Hope Professor) and Dr M. Scoble (Acting Curator until January 1985) for reading the manuscript and offering constructive comments; also Mr I. Lansbury and Mr C. O'Toole for assistance with some queries which arose regarding the Collections.

To the Pantin Trust for a grant of £4500 towards publication costs; this was obtained through the good offices of Professor A. G. M. Weddell, former Chairman of the Hope Curators, and I would also like to thank him for his continuing interest in the project.

To Committee for the Scientific Collections in the University Museum for a grant of £500.

To Visitors of the Ashmolean Museum for permission to reproduce portraits of the Reverend F. W. Hope, Mrs Hope, and Professor J. O. Westwood.

To the Keeper of the Archives, University of Oxford, and Miss Ruth Vyse for help given when consulting the archives.

To Miss Denise Blagden (Scientific Collections) for photographic work.

To Richard Hanage, grandson of K. R. Hanitsch, for access to 'History of the family Hanitsch'.

Lastly, a special thank-you to Ernest Taylor who started it all and for so many years gave devoted service to the Department and Collections until his retirement in 1979.

Having completed this work I should like to end by echoing the words of Professor G. D. Hale Carpenter: *Finis coronat opus* (the end crowns the work).

Contents

Plates

Plates fall between pages 50 and 51

1. Frederick William Hope and the Deed of Gift

T HE Hope Collections owe their origin to the Deed of Gift executed by the Reverend Frederick William Hope 'of Upper Seymour Street, Portman Square, in the County of Middlesex and late of Christ Church, Oxford', on 4 August 1849, which made over the whole of his valuable entomological collection, library of natural history, plates, engravings, and other articles to the University of Oxford.

Frederick William Hope (Plate 1) was born on 3 January 1797 at 37 Upper Seymour Street, London. He was the second son of John Thomas Hope of Netley Hall, Shrewsbury, and Ellen Hester Mary, only child and heiress of Sir Thomas Edwardes, Bart. At first he was privately educated under the tutorship of the Reverend Delafosse of Richmond. In January 1817 he came up to Christ Church where he remained until 1820. He was given the usual classical education of the period and graduated as BA in December 1820 and MA in April 1823. He was then presented to the curacy of the family living of Frodesley in Shropshire, but his health did not allow him to continue the vocation for which he was so eminently suited.

During his residence at Oxford he came under the influence of Dr John Kidd[1] whose lectures on zoological subjects made a great impression upon him. They became great friends, and such was his regard for Dr Kidd that he later presented a portrait of him, taken from the only likeness possessed by the family, to the new University Museum. Leisure hours were spent studying geology and other branches of natural science, although entomology was his chief interest and he found opportunities for collecting at Shotover, Bossom's House on Port Meadow, and Wytham. His interleaved copy of T. Marsham's *Entomologia Britannica* (1802) contains details of the species of beetles captured at Oxford whilst he was an undergraduate. His donations to Oxford started long before the original Deed of Gift—in 1833 he added to the collection of the Anatomy School, which was then at Christ Church, and in 1848 he made a magnificent contribution to that section of the Radcliffe Library concerned with volcanoes.

The Reverend Lansdown Guilding wrote from St Vincent to Hope on 26 August 1830:

How happy should I have been had we resided at Oxon together. In my day 1817 I had no companion but Lyell[2] of Exeter with whom I used to ramble and collect insects. Are there many young naturalists at Oxon now? Duncan I did not hear of when I was at College and the Ashmolean was in a rascally state. I knew most of the Professors from whom I rec'd the greatest attentions & had the Radcliffe at my command—alas! What would I give for that golden opportunity of acquiring know-ledge now. I would ride all night to spend 2 hours a day in that matchless Collection of Zool. Works.

Guilding's Cabinet was amongst those listed in the Schedule of Deed of Gift.

On 6 June 1835 Hope married Ellen Meredith (Plate 2) at Marylebone Church, London. She was the younger daughter and co-heiress of George Meredith of Nottingham Place, Marylebone, and Berrington Court, Worcestershire. The Merediths were close friends of the Disraeli family and in May 1833 Benjamin Disraeli had proposed to Ellen, but she refused him. Sarah Disraeli urged him not to give up and to try again, but to no avail—Ellen married Hope. There were no children.

Throughout their marriage, Hope had the full support of his wife in all his activities. Both had private means and used these freely to amass the great collections of all branches of natural science which formed Hope's private 'Museum' in Upper Seymour Street, and later to augment the collections at Oxford. In addition to the many purchases, Hope acquired material by gift and exchange. Specimens were sent by C. D. E. Fortnum, R. Fortune, W. S. MacLeay, and J. S. Roe to name but a few. One exchange arrangement was with the Reverend S. H. De Saram, Sinhalese Colonial Chaplain at Colombo. In return for specimens of all kinds (insects, snakes, shells, ivory images, gem stones, etc.), fashions and domestic items were sent for Mrs De Saram, and a handsome present of cutlery—the pride of Mr De Saram for 'Such cutlery never came to Ceylon before'. Unfortunately one of the consignments dispatched to Hope was lost at sea, the boat having been attacked by pirates near Ascension Island and the case thrown overboard. These same pirates were eventually shipwrecked near Cadiz and put in prison, having committed a 'horrible crime' on a vessel called the *Morning Star*.

Hope was a generous man. Always willing to impart knowledge to those who sought his help in the pursuit of natural history, he helped with specimens and books and on some occasions gave financial assistance when hardship was brought to his notice. His London residence was a favourite meeting place for naturalists, and his 'Museum' was open on certain days for the benefit of serious students of entomology. The *conversazioni* were splendid occasions—his collections being the envy of many and a delight to all who attended these gatherings. Whenever Charles Darwin was in London, he would call upon Hope. They spent a considerable time together and in June 1829 made a memorable trip through North Wales 'entomologizing'. Darwin called him 'my father in Entomology' and during the voyage of the *Beagle* wrote an interesting letter on 1 November 1833 from Buenos Aires:

I have many times, since leaving England, intended writing to you,—but as many times put it off.—I believe the chief cause has been a conscience not quite free from shame.—I am not the worthy slayer of sufficient Hecatombs to venture to write to my old Instructor. When I last saw you in London my promises were great, my performance I grieve to say does not equal them.—The Beagle for the last year has been cruising either amongst the islands of Tierra del Fuego or on the barren coast of Patagonia—Both these regions are most singularly unfavourable to the insect world. In Tierra del [Fuego] I *captured several* alpine Carabidous beetles, & one Carabus: & on the sandy deserts of the latter country, there are many of the Heteromera.—But there in absolute numbers are not to be compared to the booty on one of your Achilles-like onsets. Before we came to these Southern regions inhospitable to Entomologists & Insects, I did pretty well amongst the Coleoptera.—I often thought of you, when sweeping the rich vegetation of the Tropics I *captured* the smaller Coleoptera by hundreds.—If, as I believe you told me, European cabinets contain few of the minuter Beetles from tropical countries, I shall bring home a very great number of undescribed species both from Brazil & the Rio Plata.—It may be a foolish fear, but I often wonder if any person will be found who will describe so many minute insects. This fear is rather a drawback in my collecting.—Excepting Coleoptera, pudet pigetque mihi, I have done scarcely anything. The incumbrance of a box & flynet is not trifling.— When I have to carry Geological tools, fire-arms, spirit-bottle for reptiles &c &c. I hope however to improve & be more diligent in this respect.—At Rio de Janeiro I took many water beetles, most exceedingly small Hydropori, Hyphidri, Hydrobii &c &c.— Also a fine species of that curious sculptured genus (I forget the name) which lives beneath stones in running water.—I was much interested by finding this; it could not fail vividly to recall some of our walks at Netley; in a like manner chacing [*sic*] Cicindela nivea amongst the burning sand hills most forcibly reminded me of the Hyb[rida] at Barmouth.—Judging from the Pamphlet you gave me & which I have found very useful, the insects of the Rio Plata are tolerably well-known.—I regret therefore the less, not having worked as hard as I could have done.—

I took the other day a fine Leionotus.—If you feel inclined I should much enjoy hearing from you. I know nothing of the scientific world of London.—The last thing I heard about you was an age since, viz, that you were on your road to Germany & that Eyton failed to accompany you.—What is Eyton doing? He must be by this time a famous naturalist. Remember me most kindly to him. I had hoped he would have by this time been wandering in some Terra incognita.—My direction is H.M.S. Beagle Valparaiso, if you will condescend to write to so recreant an Entomologist. I shall be very much obliged. Remember me to the few friends I have amongst the naturalists & Believe me dear Hope,—Your most obliged disciple

Chas Darwin

Remember me most kindly to all your family & my congratulations (although they arrive rather late) to your brother.—Floreat Entomologia.

I shall much enjoy some scientifico-Entomologico-Gossip.–One more farewell.

It is interesting to note that in 1843 Hope offered two prizes of 5 guineas each for essays on two entomological subjects: the best compendium of the works of entomologists in this country, and the best memoir on the insects injurious to market gardeners. However, only one of these was competed for (see Newport's address to the Entomological Society of London, February 1845), and the prize was awarded to William Frederick Evans, MES, for

A bibliographical account of the works and writings on entomology; published in this country since 1574. The manuscript of this essay is contained in the Archives (see EVANS, W. F., in Appendix A, this volume).

Hope's success as a collector of English and exotic insects was considerable, and he readily contributed information and specimens for publications now regarded as classics of the day, such as J. L. C. Gravenhorst, *Ichneumonologia Europaea* (1829); W. E. Shuckard, *Essay on the indigenous fossorial Hymenoptera* (1837); H. Gory and A. Percheron, *Monographie des Cétoines* (1833) (Hope Library has his personal copy from the authors); C. J. Schönherr, *Genera et species Curculionidum* (1833–45); J. F. Stephens, *Illustrations of British entomology* (1827–45). His entomological interests were varied and included the economic aspects of silk, parasitism of insects in man (he corresponded on this subject with George Newport and Richard Owen), and the insects mentioned in the Holy Scriptures. He contemplated writing a work entitled 'Entomologia Sacra' and to this end amassed copious notes, but unfortunately it was never completed.

As his collection was world-wide, foreign specimens occasionally became mixed with the British and were then recorded from Britain in error—hence the following witty passage which appeared in the *Entomologist's Weekly Intelligencer* in 1857:[3]

Shropshire,—that *terra incognita* to the mass of entomologists,—we need scarcely call attention to, since it has been rendered immortal long ago through the captures of one of our most outstanding collectors. The only question that arises is, what does *not* occur there? *Pachyta Lamed* (inhabiting the summits of the Alps and Pyrenees), *Apate capucinus* (of Central and Southern Europe), *Lebia hæmorrhoidalis* and *turcica* (so prized even on the Continent), *Lixus Ascanii, Lepyrus Colon*, and other species innumerable, known nowhere else within the British Isles, all, all are 'Common'; and, on one occasion, it has been stated (though we cannot vouch for this personally) that they were actually found ready pinned; so let the Coleopterist who would turn his time to good account fly instantly to Shropshire, and *hope* for the best,—*spes nunquam fallit* [Hope never fails].

Occasional mistakes Hope may have made, but on the whole he was an accurate and careful collector. He is also said to have reared great numbers of Lepidoptera in cages with a layer of earth covered with damp moss.

Hope was elected Fellow of the Royal Society in June 1834 and of the Linnean Society in 1822, and took an active part in the formation of the Zoological Society in 1826; he was a founder-member of the Entomological Society of London in 1833—hardly surprising, since he was in touch with most of the professional and amateur entomologists of his day. He served as its first Treasurer and was President in 1835–6, 1839–40, and 1845–6, and Vice-President in 1833–4, 1837–8, and 1841–3. He regularly attended meetings of the Entomological Society, until ill-health led to his resignation in 1847. In 1833 he was elected a Member of the Société Entomologique de France.

Many honours were bestowed upon him by learned societies:

1832 The Naval and Military Library and Museum, London.
1836 Real Academia de Ciencias Naturales y Artes de Barcelona.
1842 Société Philomatique de Paris.
1844 Académie Royale des Sciences, Turin.
1846 Société Linnéenne de Lyon.
1847 Silver Medal presented by the Grand Duke of Tuscany (Leopold II) in acknowledgement of a collection of birds presented to Museo de Fisica e Storia Naturale di Firenze.
1848 Imperiale e Reale Accademia Economico-Agraria dei Georgofili de Firenze.
1848 Institut d'Afrique.
1849 Münchener Verein für Naturkunde.

The first stone of the new Museum at Oxford was laid in 1855 and on this occasion the degree of Doctor of Civil Law was conferred on Hope by the University—his collection having been a major factor influencing the decision to build the Museum.

In spite of his delicate constitution Hope travelled extensively on the Continent in the 1830s (France, Switzerland, Germany, Holland) but by 1844 his health was giving cause for concern and from 1849 he was spending long periods abroad in a warmer climate, his choice being Naples and Nice. In spite of his suffering, which was further aggravated by a 'virulent ague' contracted in Holland, and an accident at Nice when he fell into the sea and almost lost his life, his interest in collecting was undiminished. The Mediterranean climate suited him admirably and he continued to acquire large collections of fishes, crustacea, shells, birds, and insects. Towards the end of his life, enforced inactivity did not prevent him from adding to his collections; he would keep up to date with sale catalogues and send a representative to purchase items on his behalf.

Hope had evidently discussed with Dr Kidd his desire to give his collection to his Alma Mater and negotiations commenced with a visit from the Reverend R. Greswell, Tutor of Worcester College, carrying a letter of introduction from Kidd. Final arrangements leading to the completion of the Deed took place during the Long Vacation in 1849 when unfortunately nearly all the members of the Hebdomadal Board were absent from Oxford; consequently the Vice-Chancellor, Dr F. C. Plumptre, Master of University College, was unable to put before the University authorities the communications from Hope via Greswell concerning the intended gift, including the draft of the Deed which had been sent for perusal. Greswell would appear to have been the chief negotiator and as a result considered he was the 'most authentic interpreter of Mr. Hope's wishes and intentions'. In a letter to Greswell, dated 11 July 1849,[4] the Vice-Chancellor informed him that he had suggested a few alterations to the Draft which appear to me desirable in order to carry out more fully Mr. Hope's intentions [and] to adhere as closely as possible to what I understand from Dr. Ackland[5] [Acland] and yourself to be the wishes of Mr. Hope. And I trust he will not see any reason to object to the suggestions which I have made. If however any Particulars have been omitted in Mr. Morrell's[6] copy, which Mr. Hope may wish to retain, I beg he will be so good as to insert his own inclinations.

Although unable to obtain the necessary authority at that time, the Vice-Chancellor had 'every reason however to anticipate that the Conditions are such as the Hebdomadal Board would readily assent to, previous to their submitting to Convocation that this valuable gift be accepted by the University'. It was also obvious from preliminary negotiations that Hope was anxious for his collections to be removed to a place of safety as soon as possible. The Vice-Chancellor, however, felt that 'no part of this collection should be removed to Oxford till after it has been formally accepted by the University in the next Michaelmas Term'. But in order to meet Hope's wishes should the necessity arise, he concluded his letter with the following statement:

I will venture to say, on behalf of the Curators of the Taylor Institution, that it may be placed for the present in one of the vacant Rooms in their Building; where it may be preserved, free from injury, till such time as a suitable place may be provided for the Collections.

The Vice-Chancellor's suggestions were obviously acceptable to Hope since the Deed of Gift was duly enacted (see Appendix C) and within a matter of days he was arranging for the Collections to be transported to Oxford. Several months then elapsed while the conditions attached to it were carefully scrutinized. Observations were made by the solicitor acting for the University in December 1849, and a report was submitted by 'the Committee on Mr. Hope's proposed Benefaction to the University' (see Appendix D). The Hebdomadal Board (see Appendix E) then felt it their duty to 'communicate with Mr. Hope on the subject of the Conditions contained in his proposed deed of Gift to the University, in explanation of the sense in which the Members of the Board interpret them, and in which alone they could recommend to the University to accept the obligation'. As a result of these deliberations, alterations to the first, second, fourth, and sixth clauses in the Deed of Gift were submitted by the Vice-Chancellor for Hope's consideration through his solicitors, who recommended that he accede to the required alterations. This he would not do and his solicitors informed the Vice-Chancellor in March 1850[7]

it is still open to the University to accept Mr. Hope's Collection according to the terms of the original deed of Gift Mr. Hope having conveyed to us no instructions to the contrary but having merely negatived the proposed modifications of the original deed. We should have written you on the subject but thought Mr. Hope had fully intimated his views to you.

Whilst these negotiations were in progress the Vice-Chancellor had also been in communication with Hope, who was by this time residing in Nice and in a poor state of health. Many strongly worded letters were sent by Hope in reply; obviously upset by the Vice-Chancellor's letter of 4 January 1850, Hope wrote in reply at the end of January,[7] stoutly defending the conditions laid down in the Deed:

Your letter contained *3 objections* to the deed of Gift . . . *I repeat* again I did not wish to dictate & certainly I ought not to be dictated to in a deed of gift . . . How my intentions have been perverted I know not. I offer my collection to Oxford & in the endeavour to have it taken care of, it is thought my intentions may lead to litigation. I have ever been a man of peace, & have yet to learn how my views have been construed so as to cause disputes. As to Mr. *Morel*'s objections, I considered them frivolous. He may be a *Morel* a *truffle* or a cerise. He may be descended from the dictionary man & may find terms probably better adapted to the University than mine. My lawyers, *relatives* & friends (& I tell you *several* gave decided opinions on the clauses) recommended the deed of gift as sent to Oxford. It was with difficulty *some* of them consented to the clause which you recommended. Any *reasonable modification* which may prevent disputes & probable litigation hereafter may be attended to, but I feel *confident* from the long delay in receiving letters from England, that other difficulties have *arisen* or are likely to be brought forward. I must however halt for the present, being far too ill to write anything more . . . *Jany 29th* To return to the charge . . . My object in giving the collection during my life time, was to save the Probate duty, & it is the first time that 'bis dat qui cito dat' ever caused me a doubt.

On 14 February he again wrote in reply to a letter from the Vice-Chancellor:[8]

Your letter has greatly surprised me. I will endeavour to answer it altho' I am suffering most acutely. . . . The Board is evidently I think unacquainted with the terms on which I offered my collections. The first thing I stipulated for was 'that my collections shd be kept together (i.e. *Cabinets, Insects, Crustacea, Mollusca, Books, Topography, Portraits & Fossil Insects & Crustacea*, the resins & Ambers &c. &c.) and *not dispersed.*' On the *first* stipulation I ask no more that what my kind Patron *Sir Joseph Banks asked* when his Books were to be transferred to the British Museum, they still remain a separate collection. 2^{ly} I wished my collections to be accessible as far as possible particularly to Foreigners—3^{ly} My collections were to be an adjunct to the Ashmolean. *From your letter it appears now*, that they *may* form a part of it, or may not. 4^{ly} My Curators are to be *Ciphers* & under the domination of Mr. Greswell all his *schemes* of classification will no doubt be carried out (Vd note at the end of this letter). You say that you *verbally* understood me that I had no objection to such arrangements. My opinion alluded *solely* to the use of the books of my library for the benefit *of the Ashmol*ean members. 5^{ly} As to evening arrangements I never contemplated them. In most Museums & libraries Candles & Fires are not allowed of an evening. 6^{ly} As to limiting the period & making it depend on the University not fulfilling its engagements, it w^{d} indeed be dictating on my part, & as I wish not to be dictated to, I certainly cannot adopt the views of the Board.—Enough has now been stated I think, to convince you that I am not likely to change my opinions. Pray convey to the Board my most respectful thanks for the trouble I have occasioned, I greatly regret then I cannot accede to their request—as the stipulated terms on which *I offered* & (now offer) the collections are not likely to be entertained. The Deed of Gift, as it is, must be either accepted or rejected . . . (*Note*) Sundry clauses were added to the Deed of Gift in consequence of the development of Mr. Greswell's plans respecting my Collections.

Meanwhile Hope had received yet another letter from the Vice-Chancellor to which he replied in March[9] to the various points raised as to the disposition of the collections, a major part of which had arrived in Oxford soon after the Deed of Gift had been signed:

With regard to your objections (or rather those of the Board) I can only repeat, that so long as the *Collections* are in transit or in the Taylor Building the Trustees must carry out their views. The sooner *they* [collections] are removed, the better it will be for Science. Secondly, 'You always understood that it was my wish that the Books shd be kept together' & I must *add* the *Crustacea Insecta* & *Arachnida*, with the Books! To convince you that I do not raise imaginary objections I am disposed to allow all the Vertebrate Animals of my collection to be added to the Ashmolean Museum & together with them the *Mollusca*, *Testacea* & all Fossils excepting Fossil Insects and Crustacea. Hitherto I have not sent any thing like a collection of *Birds*, *Fishes* or *Reptiles*, altho' I possess many in each of those departments. My *Library*, *Portraits*, *Topography*, *Granger*, *Icones* &c ought to be in the same room as my Invertebrata. Such an opinion I stated to Mr. Greswell. *He thinks* differently now with regard to my *Curators*. I wished them to be unshackled, & in case of endowment (which may naturally be expected) should the Museum be built. I think they ought not to be under the control of other *Curators* who take up *other* departments. I am afraid that *Vertebrate* & *Invertebrate* naturalists cannot accord. Subdivision of labor in the present rapid increase of science is absolutely necessary. As an argument in favor of my views, I may mention that Linnaeus shortly after publishing his Systema Naturae gave up to Fabricius the Crustacea and Insecta! Cuvier also gave up the same Classes &c. to Latreille, & the opinion of Cuvier is not one to be lightly disregarded. As to mere household arrangements which you call general regulations, I am not likely to object to them.—Your letter intimates that *I* or *my successors* may demand back the collection for minor matters. Now allow me to ask you Is it likely that a collection w[h] has employed me 30 years, & cost me very considerable expense, will be demanded back, when I have no longer sufficient energy to keep it up. It has outgrown my early expectations greatly. Did I receive it back, what am I to do with it? Sell it, no, never. Sooner would I burn it!—As to my *Successors*, so long as the Collections are kept together according to my wishes, they are never likely to demand it back for any minor points. To conclude I do not like Mr. Morell's document, the wording of it displeases me—The Board evidently cannot trust me & I therefore see no reason why I should trust them. I must *repeat* & *repeat* again & again. The views & intentions of a Donor should not be doubted. Consult my friend *Kidd*. He believes me I think honest & if you cannot arrange matters, I cannot help it . . . PS. Pray do not write again. I am far too ill to attend to anything. This letter has exhausted my strength. I am ordered up to the Hills for a change of air.

In spite of the misunderstandings that had arisen on either side, wiser counsels prevailed. The original Deed of Gift was accepted by Convocation in April 1850 and in May the Hebdomadal Board[10] agreed that a letter of thanks should be proposed in Convocation and be sent to Mr Hope. Accordingly the following letter was sent:

To the Reverend Frederick William Hope M.A.

We the Chancellor Masters and Scholars of the University of Oxford request you to accept our public acknowledgement of your recent munificent Donation.

We are duly sensible of so distinguished a proof of your attachment to our University. We indulge the hope that this rare and valuable Entomological Collection may tend, in a great degree, to encourage among us a more extended cultivation of Natural History. And we recognise a peculiar claim upon our Gratitude in your having afforded this Example of your munificence at a time when a more than ordinary desire is manifested to attract the attention of our youth to the study of the Physical Sciences.

Given at our House of Convocation, under our Common Seal, this fourteenth day of May in the year of our Lord 1850.

Hope's generosity did not end with the original gift. Both he and Mrs Hope, who shared his interest in natural history and was the first lady Fellow of the Entomological Society of London (elected 1835), continued to augment the collections of specimens, books, and portraits for many years. His concern lest the results of a lifetime's work might not be given the proper accommodation and care they deserved often led to misunderstanding between him and the members of the University actively concerned with the benefaction. John Obadiah Westwood, a friend of Hope's for many years, who had an intimate knowledge of his collection, appeared to act as a mediator in these matters and also kept a watchful eye on the collections in general. Although living at Hammersmith, he made frequent visits to Oxford and was much involved with the arrangements. Hope evidently relied implicitly on his judgement. It was no surprise therefore that he pressed for Westwood to be appointed 'Conservator' of his collections and ultimately nominated him as the first Hope Professor of Zoology, a well-merited position.

Notes and References

1. John Kidd (1775–1851), Aldrichian Professor of Chemistry 1803–22, Dr Lee's Reader in Anatomy 1816–44, Regius Professor of Medicine and Aldrichian Praelector of Anatomy 1822–51, Radcliffe Librarian 1834–51.

2. Charles Lyell (1797–1875). He was originally interested in entomology but was attracted to geology through the lectures of Dr Buckland. His book *Principles of geology* influenced Darwin in the development of his theory of evolution.

3. *Entomologist's Weekly Intelligencer*, 2 (1857), editorial comment, pp. 118–19.

4. Oxford University Archives, W.P. α/40/15, f. 1.

5. Henry Wentworth Acland (1815–1900), Fellow of All Souls 1840–7, Dr Lee's Reader in Anatomy 1845–57, Regius Professor of Medicine 1857–94.

6. Frederick Joseph Morrell was a well-known Oxford solicitor and Steward of St John's College, who was also appointed solicitor to the University in 1854.

7. Oxford University Archives, W.P. α/40/15, ff. 3–4.

8. Oxford University Archives, W.P. α/40/15, ff. 5–6.

9. Oxford University Archives, W.P. α/40/15, ff. 9–10.

10. Oxford University Archives, W.P. γ/24/6, pp. 188–9.

2. Hope Curators and the appointment of Conservator

UNDER the 1849 Deed of Gift Hope stipulated that the Vice-Chancellor, the two Proctors, the Regius Professor of Medicine, and the Keeper of the (old) Ashmolean Museum and their successors, plus two other *ex officio* members, should be appointed Curators of his Collections. They had full powers and authority to frame rules and regulations for the safe custody and preservation of the Hope Collections, and were known as 'Mr Hope's Curators'. Soon after the formation of this body, Dr H. W. Acland was requested to act as 'Honorary Conservator' of Hope's Collections until such time as an official appointment was made.

The first meeting of the Hope Curators recorded in the Minute Book took place in March 1857 when they dealt with the proposed location of Hope's collection in the new University Museum and his wish to supplement the salary of Westwood if he were appointed Conservator.

Charles Drury Edward Fortnum, Hon. DCL (Oxon), an entomologist and close friend of Hope, who also donated a large number of works of art to the University, played an important part in negotiations with leading members of the University for the acceptance and proper housing of the Collections, and to ensure that Hope's wishes were complied with. On 11 March 1857 he wrote to Westwood:

... On the 17th Feby six delegates were appointed for the New Museum. These have no *direct* connection with the Hope Coll: more than as it will be incorporated with the general new Natural History Museum. . . . Mr. Hope's object is to have a scientific Entomologist as the Curator of his collection and without doubt a coll: of that importance to the Science ought to be in the care of one who would develop it in the manner it deserves & not merely in the keeping of an inexperienced young person. The only interest I have in the matter is to see the collection properly cared for & Mr. Hope's wishes carried out as they ought to be by the University. . . .

On 24 March 1857 Westwood wrote from Hammersmith to the Delegates of the Natural History Museum (new University Museum) applying for the 'Curatorship of the Entomological portion of the Hope[i]an Collections', and in support of his application he quoted a letter received from Hope dated 11 March 1857:

. . . as you have known my collections above 20 years & have arranged the greater part of the Insects &c. &c., I wish you particularly to be my Curator, as you know the original specimens of *Lee's* Cabinet *named by Fabricius* . . . The geographical

arrangement of my Insecta cannot be known by anyone but yourself. I do not wish it
to be disturbed as it is the result of much labor.

At a meeting of the Hope Curators on 22 May it was resolved:

That Mr. John Westwood be appointed Conservator or Managing Curator or Inspector
of the Revd. Frederick W. Hope's Entomological Collection for the term of five years
from the present date at a salary of Fifty Pounds per annum; the said Revd. F. W.
Hope having undertaken and agreed to pay a further sum of Two Hundred Pounds
per annum to Mr. Westwood for his services as such Conservator during the said
term of five years.

Dr Acland then resigned as Honorary Conservator in October and the Hope
Curators authorized his 'formal delivery of the keys to Mr. Westwood'. At
this meeting:

A letter was read from Mr. Westwood communicating large & valuable additions
made by Mr. Hope to his former Donations, including Mr. Westwood's Entomological
Collection, which had been purchased by Mr. Hope, & the donation of many valuable
& rare Insects, Books, &c. by Mr. Westwood himself.

Subsequent meetings were held at irregular intervals and were concerned
with the administration (curatorial and financial) of the whole of the Hope
Collections.

In 1896 Professor Poulton, who had succeeded Westwood in 1893, sought
the opinion of Mr G. A. Boulenger, who was a First Class Assistant at the
British Museum (Natural History), regarding the condition of the collections
of dried reptiles, amphibia, and fishes. Unfortunately Boulenger considered
the great majority to be worthless. This was reported to the Hope Curators
at their meeting in May and 'The Curators after seeing the specimens recommended that they should be destroyed as soon as possible, together with
other worthless specimens of birds & mammals also stored in the lofts'.

In December 1899 they agreed in principle that books and pamphlets not
directly of use to the Hope Department might be disposed of for the benefit of
both the Collections and the Library. As a result many valuable works and
old cabinets have been sold over the years. Recently much attention has been
given to the restoration of fine cabinets and valuable books in the Library.
Three items of furniture deserve special mention: first, the satinwood cabinet
listed in the original Deed of Gift which formerly belonged to Johann Jakob
Dillenius (1687–1747), first Sherardian Professor of Botany at Oxford. This
cabinet originally contained fossils, shells, minerals, etc. and (by 1857) a set
of aphids mounted in Canada balsam by Mr F. Walker. The aphids were
transferred to the main Collections when the cabinet with its remaining
contents was 'Transferred to Rowell's[1] charge in Zoological Room' at the
(old) Ashmolean Museum (pencilled note by J. O. Westwood). The cabinet

was eventually brought back to the new Museum (noted in Westwood's Report for 1879), where it would have been cared for by Westwood, but with the passing of time it was relegated to the loft area where it remained until 1977. As a result of enquiries at the Victoria and Albert Museum it was realized that it was a rare relic from the late seventeenth century. The cabinet has now been restored and stands in the Hope Library.

The second important cabinet is the one now containing Westwood's specimens illustrating 'Economic Entomology'. The original inlaid cabinet was given to Miss Emma Swann, his niece, in 1895 with the approval of the Hope Curators, the twenty-six small drawers being retained and placed elsewhere. These have now been brought together and assembled in a cabinet specially made for them, a brass plate indicating the contents. This cabinet stands in what is now the Historic Room.

The third item is a handsome early Victorian side-table. For years it stood in the gallery of the Museum covered by a table-case and plenty of dust. Now it has been restored and has pride of place in the Hope Library.

One particularly interesting sale was authorized by the Hope Curators, and took place in March 1903 at Stevens, King Street, London, when the following items were auctioned:

Part of a beam of the Royal George recovered by Col. C. W. Pasley in 1839, fetched 12 Guineas, and a bottle of soda water, half full, but with cork and wire fastening intact, raised from the same ship, realised $25\frac{1}{2}$ Guineas.[2]

The *Royal George*—pride of the Navy and oldest first-rate man-of-war in the service—sank at Spithead in August 1782. A victualling sloop called the *Lark*, laden with rum, was lashed alongside. The rum was put on board on the side of the ship already low in the water. This, together with the men so employed, caused the *Royal George* to list even further, and soon she was taking in water and eventually sank, pulling the sloop with her. Out of 1200 people on board, including 250 women and children, nearly 900 lost their lives—an event paralleled only by the loss of the *Mary Rose* in 1545.

According to the minutes of meetings held from time to time, the Hope Curators exercised control over the engraved portraits and topographical prints and also elected the Keeper of the Hope Collection of Engravings. This continued until about 1934 when responsibility for the portraits and prints was transferred to the Ashmolean Museum. The Hope Curators, however, still met and discharged their obligations in connection with the Entomological Collections and Library, as indeed they continue to do up to the present day.

NOTES AND REFERENCES

1. George Augustus Rowell, an Assistant Keeper at the Ashmolean Museum, who later held a similar appointment at the new Museum.
2. Reported in *The Times*, 11 March 1903.

3. New University Museum and endowment of Hope Chair

By 1856 Hope was actively taking steps to found a new chair in the University to be known as the 'Hope Chair of Zoology'. He also gave much thought to the new Museum, which was in course of being built, and wrote to Acland from Hotel d'Europe, Rue Piedmont (letter undated) suggesting

the formation of a provisional Committee of the Zoological departments, say

Vertebrata	Mammalia	—	Dr. Acland
	Birds	—	Strickland
	Fish &c &c	—	Yarrell
Invertebrata	Crustacea	—	Hope
	Insecta	—	Westwood
	Vermes &c	—	Gray
	Mollusca	—	Grant &c &c

As you are at the Head of the Nat. Histy Department, only suggest for Zoology *two* provisional Committees of *Vertebrata* and *Invertebrata*, a report of both may be given annually by yourself & tend much to encrease [*sic*] the collections & convince the unthinking that the Oxford University Museum is not very likely to go to sleep. Let *members* of any scientific Society be added to the above names (foreign as well as British) provided they are likely to take an interest in any branch.

Not until 1954 did this interesting suggestion fully materialize, with the formation of the present Committee for the Scientific Collections in the University Museum. This step was reinforced in March 1955 when the University conferred the title of Curator on the member of the academic staff in each department who had responsibility for the Collections.

Meanwhile the Hebdomadal Council directed Dr H. W. Acland, Honorary Conservator of Mr Hope's Collection, and Professor J. Phillips, Keeper of the Ashmolean Museum, to report 'on the steps it may be desirable to take for preparing the *Hope Collection* to be transferred to the Museum'. It seems that this Report, dated 23 May 1856, led to the eventual dispersal of Hope's collections from the Taylor Institution. He wanted all his Collections, Library, and Engravings to be kept together, and was under the impression that it had been tacitly agreed to house all these in the new Museum, the erection of which had obviously been hastened by the vast amount of material donated by him. From the correspondence it became clear that a misunderstanding had arisen on this point and that in fact the University was proposing to house only the natural history collections and library in the Museum. This apparent change of policy annoyed Hope, who suspected Acland of having promoted opposition to his wishes in Convocation. At one stage the question of the assignment of rooms seemed to be the only thing standing in the way of the Professorship, and this controversy continued until the latter part of 1860. Numerous interesting letters flowed between Hope, the Reverend Edward Higgins (a relative by marriage), Westwood, and other acquaintances, whilst within University circles it caused a great deal of lobbying by interested parties. In a letter from Nice dated 27 February 1857 to the President of the newly formed University Entomological Society, the Reverend H. Adair Pickard, MA, of Christ Church, declining an invitation to join the Society, Hope expressed his irritation as follows:

. . . Oxford has ever been slow in its movements and has taken 4 months to consider about electing Westwood—no Westwood no endowment, I did think fifteen or twenty thousand pounds might be a Godsend to the miserable University. In future Oxford has no reason to expect anything from me. . . .

In spite of this assertion Hope still maintained a lively interest in affairs at Oxford and during the first half of 1857 (March to May) there was a lengthy correspondence between him, his lawyer, and the then Vice-Chancellor (Dr D. Williams, Warden of New College) regarding suitable accommodation in the new Museum for his Collections and Library. In a letter to the Vice-Chancellor dated 16 March, Hope wrote:

I beg leave to thank you for your kind letter of March the 10th. The major part of it is very satisfactory, particularly what relates to the Portraits and the Books intended for the Taylor Buildings. Mr. Greswell's letter informs me that the *Rooms intended for me* are destined for a Library, and offers a room 37 ft. by 24 and a Curator's Room which I think little of. There is not a doubt in my mind that *adequate space* ought to [be] granted for my Collections. Mr. Greswell informs me also that my Library is to be handed over to the New Library [the original Radcliffe Library]. *It is contrary to the Deed of Gift* . . . I expect ample room for present or future collections of Crustacea and Insecta. I cannot allow my library to be swamped by another. It will be under the same roof, and always I hope accessible. My dear Sir, I have made I think *a liberal offer*. I did not expect such changes and deviations. I will write to my Lawyer again. I ask nothing unreasonable. If my wishes are not attended to I withdraw my proposal. . . .

Hope had just cause to be annoyed. His generosity knew no bounds—he had even agreed to pay for the furnishings of the rooms allocated to his Collections in the new University Museum. By the end of April, however, agreement was reached to the satisfaction of Hope, the Hope Curators, and the Delegates of the new University Museum. The Entomological Collections were to be placed in the South Room and the Library in the private room adjoining. The solanders and books of natural history dealing with quadrupeds, birds, fishes, reptiles, Mollusca, and Testacea were to be placed in a bay in the great West Room and kept as a separate entity. This was the room Hope thought had been assigned for the whole of his natural history collections but had in fact been allocated to the 'New Library'. There they remained until 1894 when they were moved to the greatly extended Hope Department. Subsequently, many of the works not directly required by the Department were distributed to the appropriate science departments, Radcliffe Science Library, or sold by order of the Hope Curators.

Dr F. C. Plumptre was also disturbed about the rooms offered to Mr Hope. In a letter to Westwood dated 2 November 1858 he wrote:

I am much obliged to you for your note, and I will name the subject to the Dean of Ch[rist]. Ch[urch]. He & myself have made ourselves responsible for the cost of colouring the ceiling of Mr. Hopes rooms, in order to save the credit of the University, as it was left in a very mean & unworthy condition. But I shall not wear any pecuniary responsibility in regard to the decoration of the walls. My own opinion is that it will be better to make the ceiling good in the first instance, & then we shall see our way better for other wall painting. I entirely agree with you, that the blue on the sides of the windows is too strong for effect, and I think may be prejudicial to the appearance of the Insects, as being too bright and overpowering. I will see Mr. Swan about this.

The ceiling had apparently been left in an untidy state by the removal of gas fittings. In the event, the walls were painted blue and enhanced the rich colours of the ceiling decoration. The upper part of the room still has this original decoration with the Magdalen lilies (in honour of Westwood) making a frieze. It must have been a fine room, with the beautifully carved stone fireplace (Plate 3). Later Hope jocularly remarked 'As to the Blue Chamber it is called I hear, Hope's Bluebeard Room . . .'.

At the point where some doubt existed as to whether the University would accept Hope's proposal for the endowment of a Professorship of Zoology, Greswell[1] wrote at length to the Vice-Chancellor (Dr F. Jeune, Master of Pembroke College) on 4 February 1860 regarding the question of the room and the Professorship:

. . . It is extremely important that no time should be lost, in bringing these questions to a final settlement—questions which have been kept pending for the long period of Eleven years, (nearly). In the first place, Mr. Hope's health, (as appears from his own letters,) is in a most precarious state; his life, at the present time, not being worth so much as six months purchase. . . .

Charles Daubeny (Professor successively of Chemistry and Botany) also wrote an interesting letter to the Vice-Chancellor dated 17 February 1860. In it he refers to the objection made by some members of Council to the proposed Chair on the grounds that the subject did not provide sufficient scope to justify such a post. Daubeny emphatically dissented from this opinion which he regarded as founded 'upon an inadequate idea of the extent which the field of Zoology at the present day embraces'. He had sought the opinions of several eminent men including Professor Richard Owen, Sir Roderic Murchison, Sir Charles Lyell, Charles Darwin, Michael Faraday, Thomas Bell, and the Reverend L. Jenyns, who were all in substantial agreement with his view. Daubeny's memorandum, incorporating the views of Mr Bell and Dr J. E. Gray, Keeper of the Zoological Collections in the British Museum, is preserved in the Hope archives. Gray's views were particularly helpful and mentioned incidentally that there were currently two Professors of Vertebrata and two of Invertebrata in the Jardin des Plantes, Paris.

During the course of the preliminary discussions by the University of Hope's proposal to endow a professorship, some other suggestions had been made which did not please Hope. One was the decision to separate his collection of portraits and engravings from the natural history collections. Regarding the former, Hope stated in his reply to a letter received from Westwood and dated 23 February 1860:

I always wanted my collection to be a national one & wishing to fill up many gaps is certainly not diminishing the collection, if Oxford will only speak out on this opinion I will soon arrange things. In short as far as I can make it out, you vote with Acland & the Vice-Chancellor that the large room [in the Museum] must be abandoned for the new intended one & that the collection of Portraits is to remain without any additions till it rots in the room in which it will be deposited. Adieu to all my dreams . . .

Another proposal displeased him—to allow the Linacre Professor[2] to have control over his natural history collections. In the same letter he continued:

Who is the Linacre Professor. I never heard of him. I wish you Westwood to add any clause which will guard my collections. Comparative Anatomists destroy many unique specimens. Any dissected specimen made by the Professor must be replaced in my collection. Rare or rather unique species ought not to be touched. Duplicates may be experimented on. Pray do not write any more I cannot bear it. I shall never reach Oxford again. It is all up.

In a letter to Westwood addressed from Nice and received 2 March 1860, Hope again referred at some length to his irritation over these proceedings and said:

You mention a startling fact *that all along* there has been an opinion that the collection of Portraits and Topography w^d be out of place in a Natural Science Museum. Why was not that opinion made known to me. I then w^d have offered the *Collection to Cambridge* where I know it would be well received and treated carefully. . . . The Vice-Chancellor appears to adopt Acland's views I am sorry for it. If I find Acland continues his vexatious opposition I will not endow a Professorship. He has behaved badly to me. I will have no more to do with him.

Gossip about events at Oxford spread far and wide—even William Wilson Saunders commented in a letter to Westwood dated 27 February 1860: 'How are you getting on at Oxford. I hope there will be no hitch about Hope's proposition. The Oxford Dons must be daft if they reject the offer.'

Meanwhile Westwood had written on this subject to Mr Higgins, who replied in a letter on 3 May 1860:

As to £10,000 not being a sufficient sum for a Professorship, altho' only one or two may have openly avowed it, it is clear from your letter & from Mr. Greswell's, that you both are of the same opinion—in as much as you suggest a moderate addition which I suppose means some thousands or at least a sum in proportion to the fabulous estimate you have formed of Mr. Hope's wealth—& Mr. Greswell in his letters never fails to hint to Mr. Hope to offer further donations. Now under the circumstances this cannot be justified. In the opinion of all I have consulted £10,000 is considered ample provision for a Professorship at the University, more particularly as in course of time the Professor will be elected from members of the University who will have other sources of emolument—Mr. Hope however in a letter from Nice to the Vice-Chancellor has withdrawn this offer of £10,000—& in that determination he has the concurrence of Mrs. Hope & his friends.

The letter referred to had been written by Hope on 3 March withdrawing the

proposed endowment of a Professorship of Invertebrated Animals & every other advantage I may have proposed for the furtherance of Art & Science in the University excepting in discharging faithfully the arrangement entered into for the part payment by myself of Mr. Westwood's salary.

However, misunderstandings were resolved and by an Indenture dated 20 December 1860 between the Reverend Frederick William Hope and the Chancellor, Masters, and Scholars of the University of Oxford, an endowment of £10 000 invested in New 3 per cent Annuities was made to found the Chair for a Hope Professor of Zoology. The Indenture, a copy of which is preserved in the Hope Library Archives, laid down that

The duty of the Hope Professor shall be to give Public Lectures & Private Instruction on Zoology with special reference to the Articulata at such times as shall be prescribed or approved by the University & also to superintend & arrange the Hope Collection of Annulose Animals and to take charge of the Natural History portion of the Hope Library.

Regarding the Entomological Collections, the duties are virtually the same today as stated in the *Statutes, Decrees and Regulations of the University of Oxford 1981* (p. 321):

The Hope Professor of Zoology (Entomology) shall undertake the charge of the Hope Collection of insecta and arachnida and of the Natural History portion of the Hope Library and shall superintend and arrange the collection.

In January 1861 Westwood was nominated by Hope to be the first Hope Professor of Zoology, and he gave his inaugural lecture in April in the Geological Lecture Room of the new University Museum (Plate 4).

The Hope Department of Zoology was effectively established with the inauguration of the Hope Professor, but the arrangements made for the accommodation of some of the natural history collections apparently did not meet with Westwood's full approval. In a letter to A. H. Haliday dated 21 January 1863 and addressed from 'University Museum' he writes:

The old Ashmolean Museum is broken up & the animals have been brought to this new fantastical building—in which you may fancy how wretched the arrangement for zoological purposes are when I tell you that Cases covering about 6 square yards are allotted for the general Collection of Mammalia! & that there is not a room where poor Strickland's Collection of birds—one of the most complete ever formed & preserved in skins–can be placed & I suppose the University will lose them in consequence. The Ashmolean Insects are come under my charge . . . & a good deal damaged by Anthreni—so much for neglect,—without proper Cabinets which I am endeavouring to provide.

I go to London regularly to attend the meetings of the Entl. Society or I should die of Ennui here—there is no sympathy for Entomology or even general Zoology among the heads of houses &c. & the professors & tutors & fellows of Colleges have something else to do & do not care for an interloper among them as I am.

This letter is preserved amongst the Haliday correspondence in the Royal Entomological Society of London.

The following remarks are gleaned from a newspaper article dated 14 August 1874, and pasted into Westwood's own copy of his *Thesaurus*, regarding the establishment and duties of the Hope Professorship:

. . . The new Professor of Entomology, a most worthy man, did all that he could to make his subject attractive and himself pleasant. He organised entomological excursions, and exerted himself to the utmost to make them at least as attractive as Nuneham picnics or croquet parties in the Deanery gardens of Christchurch. He gave lectures of a most amusing kind, which were as good in their way as University sermons, and far more practical. And, in addition, he did what many professors neglect to do—devoted himself to his office, was always to be found at his post, and worked his hardest to found a complete collection of the insect world, as a sort of *corpus doctrinae* for his successors. Unfortunately, at the very time that the Hope Professor of Zoology was first appointed, Oxford was a little touchy and sore on the point of professorships and their endowments. It had just had inflicted on it, at the cost of the suppression of five full fellowships of All Souls College, the CHICHELE Professorship of Modern History; and, guided by an experience thus bitter, it was not unnatural that the munificent bequest of Mr. HOPE should be regarded from a somewhat sceptical and adverse point of view. There flourished about that time a certain Professor of Logic [Professor H. Wall], who was accepted by common consent as the representative and mouthpiece of the more conservative and matured opinions of the University. This learned doctor denounced the new HOPE Professorship as a 'professorship of bugs'. He declared that there were only twenty kinds of the *genus* known to the civilised world—to wit, the London, the Brighton, the Oxford, the Worcester College, and sixteen other inferior sorts; that all a professor under the HOPE bequest had to do was to get twenty pill-boxes, each appropriately labelled, and to put a specimen of the unpleasant insect into each, and that he would then be fairly started in business for any lectures which he might have to give; and, finally he expressed his emphatic conviction that the HOPE, unlike the CHICHELE Professorship of Modern History, was apt to lead the youthful mind astray into the heterodox regions of natural science, and so to subvert those revealed and established truths which it was, at any cost, the primary duty of the Universities to maintain and uphold. . . .

Hope died aged 65 on 15 April 1862 and in July of the same year his widow and sole executrix fulfilled a wish expressed by her husband in his lifetime by providing a further endowment of £10 000 in New 3 per cent Annuities, the dividends being allocated thus:

. . . one third shall be paid as a stipend to the Keeper of the Hope Collection of Engravings for the time being one third to the Hope Professor of Zoology for the time being in augmentation of his present income & the remaining third shall be paid to the Hope Curators or any two of them & shall be applied as to one moiety in keeping up & increasing the said Collection of Portraits & as to the other moiety in keeping up & increasing the said Entomological Collection.

In December 1864 Mrs Hope gave a further sum of £1666. 13*s.* 4*d.* in the same stock to augment the stipend of the Keeper of Engravings, for the purpose of enabling him to employ an Assistant, and to meet expenses incidental to his duties. In January 1864 she presented a portrait of her late husband painted in oils by Lowes Dickinson, to be placed in the room containing his Collections at the University Museum. The portrait remained in that room until September 1972 when it was hung in the new Reading Room of the Hopeian Library of Entomology which had been established in more suitable rooms on the upper floor of the north side of the Museum. Later Mrs Hope anticipated a bequest made in her will and in 1866 presented a miniature of her husband to the Collection of Engravings on the understanding that 'my wish will be carried out of its remaining inseparable from the collection'.[3] Portraits (see Plates 1 and 2) of both Hope and his wife now hang in the Hope Room at the Ashmolean Museum. Mrs Hope retained her interest in the Department, and especially in the regulations governing the Hope Professorship, until her death in 1879. That year the Oxford University Commission, then in session, proposed certain alterations to the regulations. Mrs Hope sent a 'remonstrance' to the University authorities against the proposed alteration. Her anxiety about the proposals centred on the greatly extended duties which were being envisaged for the Hope Professor and which would restrict the time available for attending to the Hope Collections to an unacceptable degree. The matter was considered of such importance that the conditions of Hope's original Deed of Gift were published as an appendix to the current volume of the Entomological Society of London's *Proceedings*.

Notes and References

1. Oxford University Archives, W.P. α/40/15, f. 13.
2. The Linacre Professorship of Anatomy and Physiology was established in 1860; the first Professor appointed was George Rolleston.
3. Letter of 26 May 1866 to J. O. Westwood.

4. History of the Collections 1849–1983

Growth of the Department and the Collections

TRANSFER of those items in the Collections enumerated in the Schedule attached to the Deed of Gift commenced immediately after Hope signed the document. Some books and prints were packed and in transit by August 1849; entomological cabinets and other items followed in quick succession. Although donations of books, portraits, and natural history specimens continued to be sent to Oxford for several years to come, a large part of the original gift was in Oxford by September 1850 and was placed in rooms in the Taylor Institution in the care of the Librarian, Mr John Macray. A Visitors' Book had been started in June 1850 when members of the University came to see the Collections. Over the following years this book became a veritable entomologist's 'Who's Who' as the fame of the Collections spread far and wide. What an impressive sight it must have been to see so many important cabinets that once belonged to naturalists whose names are revered the world over by historians of science! Unfortunately, today it would be very difficult to identify any of them because of subsequent amalgamation and rearrangement, the old cabinets being dispersed when new standard cabinets were purchased.

The insect collection and other 'articulated animals' remained in the Taylor Institution together with the library, whilst the vertebrates and other animals, either dried or preserved in spirit, were transferred to the Ashmolean Museum in Broad Street in 1858. Eventually all these collections and the library were moved to the new University Museum in 1861. In the same year the collection of engraved portraits and topographical illustrations (estimated to be about 250 000), together with the relevant books, were placed in the gallery of the Bodleian (Radcliffe) Camera and remained there until 1891 when they were moved to the School of Natural Philosophy in the Quadrangle of the Bodleian Library. In September 1924 they were placed in the new Ashmolean Museum in Beaumont Street where permanent accommodation had been provided.

The Deed of Gift marked only the beginning of the Collections. By January 1851 Hope was in Naples and large consignments of specimens from the Mediterranean area were being sent to Oxford. He was also on friendly terms with Stefano Delle Chiaje, author of a learned work *Miscellanea anatomico-patologica* (1847), and in a letter to Dr Acland, Honorary Conservator of his collections, Hope writes

Great as he is in anatomy he is a child in many respects & knows little of the world,

& consequently has suffered accordingly. His enthusiasm is unabated & when his eye on looking over my Mollusca settles on a new form it is a treat to see & hear him. I do hope to have most of my Mollusca named by him, & I have therefore purchased a magnificent array of glass Bottles ready for Feby & March. In Crustacea I have filled already *16* Boxes. There are *14* large bottles containing Fish, Mollusca & Crustacea in spirits, & about 60 small glass bottles containing various branches of Natural History. My wish is to see them all arranged at Oxford. From Sicily I have received some fine Crustaceous forms from Professor Cocco & anxiously am looking for an arrival from Rizza of Syracuse.

Another notable acquisition came in May 1852 (letter to Acland). Hope purchased from a 'gentleman amateur' a fine collection of twenty-four casts for £4, Oxford paying for packing and carriage to the Taylor Building. Amongst these was 'Melpomene (large & very fine). . . Jupiter (Head very fine). . . Nelson (good). . . Socrates (tolerable). . . and Hercules (very fine)'.

Before leaving for Nice in January 1854 he sent to Acland his '*Microscopic Crustacea* most of them new to Science. Pray do not allow too strong spirit to be used, they readily decompose'. At the same time he sent an original portrait of Linnaeus: 'It would be one portrait for the Museum'. There are actually two oil paintings of Linnaeus—one of which is hanging in the Library. Later on, he also gave a fine statue of Linnaeus to the Museum, where it stands to this day. Details of the sculpture, portraying the great man in his twenty-fifth year, are contained in the Archives.

It was in 1855 that Hope began to get rather concerned about his collections at Oxford and wrote [to Acland] thus:

I am of opinion that a Curator to look after my collections is the first thing requisite. You may have Cabinets any day when the Insects are perished & nought but pins & dust remain—It is now nearly 6 years since any thing has been done to them, except adding Camphor w^h occasionally creates a mould & then turns to crystals which injure the Insects—Such I find to be the case.

A Curator ought to be appointed for my Collections—Oxford seems unwilling to do anything for *my* collections although much has been done for the Bofinees [sic] Collections (given after mine) & why? because an influential person was resident in Oxford. Why has the University prevented me having a Curator? Oxford ought to attend to my wishes. If not, I from this moment decline doing anything more. When progress in arrangement is made I may attend to things & add to the Cabinets. Ought nearly *1000* drawers to be given up? Each drawer will cost a *guinea*, money is not thought of [at] Oxford!!!—To make any promises would be deliberate folly & as to *endowment* there seems now no security for a *donor* to expect that *his* views can ever be carried out. Vested rights are thrown overboard at Oxford. For ought I know some Ignoramus may recommend my collections to go to the hammer. I would sooner see them *destroyed*. What is to be done with the Saurians? What with the skeletons of Fish? With many other things? I have not many days to remain in England. I shall leave it with all my hopes blighted & no thanks to Oxford.

It would seem as a result of this letter Dr Acland started negotiations with William Dallas[1] in May 1856 with a view to his arranging the insect collections on a piece-work basis. Dallas suggested a remuneration of £15 per month, working alternate months in Oxford and London. This would enable him to settle many points of doubt 'which are sure to occur in working up a collection. . . . Nothing would give me more pleasure than to be connected with your Museum'. He did visit Oxford as there is evidence of his using the collection, judging by the queries he sent from London to Dr Acland. But the arrangement, if it ever operated, did not last. Westwood was appointed Conservator in 1857.

On taking up his appointment as Conservator of the Hope Collections Westwood began his *Journal* (which is divided into four sections) in July 1857 with a 'List of the Cabinets and Boxes of Entomological subjects, forming the Hopeian Collection, with notes of their present condition'. This list is very detailed and forms a precise inventory of the position at that time.

In a letter written to Mr Fortnum on May Day 1857, Westwood informed him that he had offered his collections to Mr Hope for £1000—and this led to the next important entry '31st July 1857. The Entomological Collections & Library, Drawings, Engravings & MSS. of J. O. Westwood purchased by the Rev⁴ F. W. Hope & added to the Hopeian Museum.' This is followed by lists of other material presented to the Collections and includes notes regarding purchases. These entries unfortunately end in 1867.

The second part of Westwood's *Journal* starts in 1862 with an 'Account of Disbursements on account of the Annual Grant of £50 (minus income tax) for additions to the Entomological Collections made by the Rev⁴ F. W. Hope under the Professorial Deed'. This is a very detailed account of payments made up to 1881 for individual specimens, small and large collections, also books. The prices paid at that time are indeed interesting.

The third part contains only one entry, a 'General Statement of the Extent & Condition of the Entomological Collection carefully revised in March 1879 when (with the exception of a few duplicate Crustacea destroyed by Anthreni) the whole was found to be in good condition'.

The fourth part is at the end of the *Journal*: 'Annual Grant for Additions to the Hope Collections made by Deed of 1862 by Mrs Hope namely the Interest of one moiety of one third of £10 000 New 3 per cents—the other moiety being applied to additions to the Hope Collection of Engraved Portraits.' These audited accounts end in 1892.

What appears to be Westwood's first report to the Hope Curators is a letter dated 28 May 1858 to Professor Phillips, Secretary to the Hope Curators and Keeper of the Ashmolean Museum, a copy of which is pasted into the *Journal*:

My dear Sir,

I beg to announce to the Curators of the Hopeian Collections various additions recently made by Mr. Hope.

At the sale of the Library of M. Jussieu the large collection of Academical 'Eloges, Biographies, Bibliographies & Notices pour servir a l'Histoire des Sciences naturelles et physiques' formed during nearly a century by the different members of that family, and amounting in No. to about 750 was purchased by Mr. Hope & added to the library. A similar Collection from the library of M. Brongniart was also purchased by him in number about 450.

At the sale of the exotic Cabinets of insects belonging to the Entomological Society I was enabled by a grant from Mr. Hope to make very considerable additions to the Collection of Insects.

A large case of objects of natural history from the South of Europe was received from Mr. Hope during the last month—containing about 130 birds amongst which are some splendid specimens of Hawks & Eagles—about 270 dried fishes from the Mediterranean & a number of Crustacea, insecta, corals starfishes &c.

Two large boxes of Books have arrived from Mr. Hope this day containing about 200 volumes among which is the great work of Piranesi 24 vols fol. These volumes are for the most part Biographical illustrated by about 3000 engraved portraits with some topographical.

During the past Spring the whole of the dried specimens of birds, reptiles, fishes, mollusca, starfishes, corals, skeletons &c which had been forwarded by Mr. Hope from time to time to the Taylor Institute were together with 386 bottles of objects in spirits (I am counting fresh nos. of specimens) by his request removed to the Ashmolean Museum. A list of these, & of other objects of the same kind previously forwarded by Mr. Hope to the Ashmolean Museum has been drawn up by Mr. Rowell & is enclosed herein—The major part of the objects in spirits have been carefully gone over & supplied with fresh methylated spirit, by Mr. Rowell, the fishes, reptiles &c. being retained at the Ashmolean Museum whilst the Annulosa in spirits have been returned to the Taylor Institute.

The two 60-drawer cabinets ordered by the Curators last autumn are promised to be sent next week. It will be quite necessary to continue the building of cabinets on an uniform-plan for several years to come, not only for the purpose of holding fresh acquisitions at present stowed in insecure boxes, but also of replacing some of the present cabinets which are not trustworthy—I have therefore to suggest that the Curators empower me to order two more cabinets of the same size & pattern as those now ordered—to be delivered previous to June 1859.

I remain
Dr Sir
Yours very truly
J.O.W.

To this letter was appended a copy of Mr Rowell's list referred to above. The two cabinets 'ordered last autumn' were received from Mr Standish in June and August 1858, price £58 and £57 respectively.

The collections attracted many gifts of specimens and Westwood himself made donations from time to time apart from numerous purchases in his capacity as Hope Professor. Amongst the many entries in his *Journal* the following is of especial interest:

1863. August & September. Purchased from Mr. Janson a selection from the Collection of Exotic Hymenoptera purchased by him at Messrs. Sotheby's at the sale of Shuckard[2] Books. This Collection of insects contained a large number of specimens borrowed in 1861 by Shuckard from the Entomological Society, the East India House & the Collections of Mr. Hope & Westwood, & was afterwards pawned by Shuckard illegally—& after many years payment of interest on the money advanced (which at length was discontinued & the insects consequently forfeited to the Pawnbroker) the Collection was sold by auction & purchased by Mr. Janson—& by him offered to the British Museum, the Curator of which selected 100 specimens at 2/4 each which I fortunately ascertained previous to the completion of the sale to the British Museum as the selection contained a great number of Mr. Hope's & my own uniques which I claimed, but as the sale had taken place in a strictly legal manner & so long a period had elapsed I was only able to obtain repossession of the Hopeian & my own types by paying for them the price offered by the British Museum. I subsequently selected a number of other specimens from the general Collection, thus obtaining in the whole

I.O.W.

39 Thynnidae	(11 from Coll Hope
	5 from Coll Westwood
	23 from general Collect[n].
16 Mutillae	(3 from Coll Hope)
15 other Mutillidae	(2 Coll. Hope, 1 Coll. Westw.
9 other Hymenoptera	(1 Coll. Hope—Trachypus unique)

9 specimens at 2/4   52 at 2/—   19 at 1/—   *7. 4. 0.*

The amalgamation of the Hope and Westwood Collections and the subsequent additions by gift, bequest, and purchase so enhanced their value that the Hope Entomological Collections today are recognized as second in importance only to those of the British Museum (Natural History). They increased in size so much that the two original rooms became totally inadequate to cope with the numerous accessions. By 1894 the rooms marked Astronomy and Geometry (professors' sitting-rooms) and the lecture room between them (Plate 5) were incorporated with the Hope Department, having been modified for the Collections and Library, lectures, and practical work. This accommodation sufficed until 1903 when it became necessary to move some of the cabinets into the corridor because of overcrowding. Then the Delegates of the Museum erected a spacious gallery for insect cabinets in the recently vacated Radcliffe Library's principal book room, the floor space being temporarily occupied by the Wykeham Professor of Physics who moved out in 1911. In the same year this room was divided by the insertion of a fireproof floor and in January 1912 the Hope Department transferred its collections of Lepidoptera to the lower portion of the principal book room: the room which Hope had originally thought to be intended for his Collection and Library, and which caused so much controversy in 1857. This transfer enabled the rest of the Collections to be completely reorganized. In 1934 the upper room (initially occupied by Pharmacology and latterly by Soil Science) was divided into two distinct areas and allocated to the Hope Department, and in March and April of that year the moth collection was moved into the outer section (subsequently known as the Moth Room). By this time entomology, both pure and applied, had begun to attract more research students and the inner area, which contained two small rooms and a laboratory, provided for a larger number than was possible before. A generous gift of glassware, simple reagents, and other apparatus, was made by Dr D. F. Chattaway, FRS, on behalf of The Queen's College, when its laboratory was closed, thus conserving the hard-pressed funds of the department. By 1954 both areas had been converted into research rooms; recently these were demolished, and the area is now used exclusively for collections.

In 1924 a ferroconcrete floor was constructed in the reading room of the old Radcliffe Library, cutting off the finely decorated upper part in the roof space, which was then allocated to the Department. By 1926 the exotic collections of insect orders other than Lepidoptera had been transferred to this new upper room, greatly relieving the congestion which once more had arisen. This room was called the 'Huxley Room' by Professor G. D. Hale Carpenter because it was part of the area where the Huxley–Wilberforce debate on Evolution took place in 1860. The room below (the original reading room of the Radcliffe Library) was at the same time designated the 'Wilberforce Room'.

During the next twenty-five years the accommodation thus acquired was consolidated. The collections increased, especially by gifts, and it became clear that further modifications were essential. In 1957 the lower portion of the principal book room of the old Radcliffe Library was divided by a mezzanine floor (replacing the gallery) to which the butterfly and moth collections were moved. The lower floor was fitted out with research rooms leaving an adjacent area for storing the European Lepidoptera, the older historic collections of British insects and the bionomic cabinets. By 1983 these collections had been moved elsewhere to provide extra working space.

Further storage space was provided in the Wilberforce Room by alterations completed in 1962. It was not realized when all these changes were made to this fine room there would be renewed interest in the famous Huxley–Wilberforce debate, that visitors would come from all over the world to see the historic setting, but would be disappointed to find that the room as such no longer existed. This room was vacated in 1983 and the Collections moved to other parts of the building.

In 1972 rooms vacated by the Department of Zoology on the north side of the upper corridor of the museum were taken over to cater for the expanding library and the increasing number of research students. The British collections were then extended into the large room previously occupied by the Hope Library whilst the adjacent smaller room now contains the more important historic collections.

Work on the Collections

In his Presidential Address to the Entomological Society of London, in January 1862, J. W. Douglas remarked, referring to the appointment of Mr Westwood to the Chair of Zoology at Oxford, 'in order to realize the full benefits of which such an institution is capable, I trust that the Professor will have so much assistance allowed him, in attending to the Museum, as to leave him free to devote the necessary time to lectures and the other higher duties of his office'. Sadly, this assistance did not materialize. Apart from his statutory duties as Hope Professor, as Conservator Westwood worked assiduously and completely singlehanded in arranging and supervising the whole of the natural history collections, as well as attending to the administration and care of the rapidly expanding library, which as he stated in May 1891 was 'a work occupying very great care & time which is being constantly carried on & requires additional skilled assistance. This I am endeavouring to obtain by the employment at my own expense, of a young man who I am training to the work.' The assistant was J. W. Shipp who only stayed in the post for $2\frac{1}{2}$ years, resigning at the end of 1893 after a bout of typhoid fever.

Although Westwood was in touch with and collaborated with most of the well-known entomologists of his day, and many visitors came to the department to use the Collections, it was not until Edward Bagnall Poulton became Hope Professor in 1893 that the Collections were extensively worked on by specialists attracted to Oxford. They gave freely of their time and also donated valuable material; some even helped financially. Poulton himself donated material and purchased collections from his own resources. He was very successful in obtaining gifts of money for cabinets and specimens from a wide range of people. Money for new cabinets was also provided by the University to cope with the growth of the Collections, and amalgamation and rehousing of the older material.

During Poulton's tenure of the Hope Chair his personal enthusiasm made the Department a world-renowned centre for the study of the living insect in its environment. His work on mimicry in the light of natural selection led to the gathering of much information on relevant aspects such as: geographical distribution; predation, including the prey of bats, birds, and insects; courtship; and the use of scents. As a result of his study of insects in general, he elaborated his classification of coloration which became the most important theme of his work, and eventually made a significant contribution to the theory of mimicry. He was one of the best popular lecturers of the day and became 'the centre of gravity of entomological research in the British Empire'.[3]

A born collector himself, he inspired others to do likewise. As a result the Hope Collections grew so vast as to become virtually an embarrassment! He would accept anything and discard nothing, so that the Department was overwhelmed with specimens of all degrees of value, unsorted, unidentified, and hard to find. Not for nothing was his middle name of Bagnall corrupted to Bag-all by his intimates. Any expert who could be tempted to come to Oxford to help make order out of chaos was welcome. Temporary assistance was also employed on one or two occasions through grants made by Magdalen College to the Delegates of the University Museum.

Poulton was a strict disciplinarian and the conditions regarding hours of work seem odd by modern standards. When Joseph Joynson Collins was engaged as assistant he wrote in the Minute Book of the Hope Curators:

Janry 30/1905.
The conditions as to hours of work agreed upon are as follows:– Joseph Collins to be in the Department & begin work at 7.30 a.m. each week-day, having had breakfast before arrival, or taking it during & without interrupting his work. An hour's interval for lunch or dinner 1.30 to 2.30 or thereabouts (1.–2, or 1.15–2.15). Work resumed in the afternoon after the hour's interval & continued till 5.30 except on Saturdays when there is no work in the afternoon. Nett result 9 hrs. per day for 5 days; 6 hours on Saturday. Total 51 hours per week.

I undertake loyally to carry out these conditions & to do my best for the University Collections in the Hope Department.

Signed   Joseph Collins
In the presence of   E. B. Poulton

Although not recorded in the Minute Book, these conditions would presumably have applied to William Holland when he was appointed in 1893 and Albert Harry Hamm in 1897, but part of Hamm's time was devoted to printing for the Delegates of the University Museum and this resulted in a complicated entry to clarify the situation:

July 23/1900. Today a question arose as to the exact point at which A. H. Hamm's time ceases to be devoted to the Delegates & is devoted to the Department, in each day. No statement was found in the minutes of the Department, no meeting of Curators having taken place at that time.

Hence after a discussion between W. Holland, A. H. Hamm & E. B. Poulton after considering the evidence desired from customary procedure, the following statement is now made as, it is believed, substantially the same as the arrangement made in 1897 when A. H. Hamm came here. The arrangement was that A. H. Hamm should work for the same number of hours as he had worked before he came and that $\frac{1}{4}$ of his time should be given to the Delegates & $\frac{3}{4}$ to the Department. He previously worked $53\frac{1}{2}$ hrs. per week or on an average $6/53\frac{1}{2} = 8\frac{11}{16}$ hrs. per day viz. just under $8\frac{3}{4}$ hrs (exactly 8h. $41\frac{1}{4}$ min.). To eliminate fractions his day was calculated at 9 hrs. and it was arranged to allow him $1\frac{1}{2}$ hrs. a week off either in each week or to be accumulated.

Each day beginning at 7.30 & allowing the Delegates to contribute their share of the $1\frac{1}{2}$ hrs off, the point would be between 9.30 and 9.45 at which their work ceased, the whole time being from 7.30 a.m. to 5.30 p.m. in each day allowing 1 hr. interval for dinner. This arrangement having been in part forgotten by E. B. Poulton it is desirable to restate and confirm it.

[Signed]   Edward B. Poulton
W. Holland
A. H. Hamm

The passing of the years saw considerable amelioration of these conditions.

On coming to live in Oxford in 1930, Geoffrey Douglas Hale Carpenter soon found a niche in the Department and helped enormously with the Collections and Library in general. When appointed Hope Professor in 1933, in addition to coping with the mass of material acquired by Poulton, he devoted much of his time to the already large collections of African Lepidoptera. Apart from his own donations, he obtained even more material from friends in various parts of Africa and the main work of the Department for many years was centred on these acquisitions. During the First World War collections made in Africa and abandoned by German collectors ultimately found a home in the Department through the good offices of (then Captain) Carpenter, Captain W. A. S. Lamborn, and the Reverend Canon K. St Aubyn Rogers when they entered what was formerly German East Africa. However, mimicry dominated his life and for teaching purposes fine displays were at one time on exhibit in the gallery. The Rhopalocera forming the nucleus of the special collection illustrating mimicry were purchased from Watkins and Doncaster in 1895, and added to by both Poulton and Carpenter.

Under Carpenter's administration strenuous efforts were made to organize the mass of material still outstanding from Poulton's tenure. Realizing that the Library, which had been practically inaccessible, was one of the first necessities, he set to work to unravel the indescribable chaos in which he found it. Apart from cabinets and Poulton's enormous desk, the entire floor space was covered with piles of journals, reprints, and books, chest high, with narrow passages about a foot wide winding about between them. The Radcliffe Science Library came to the rescue and donated twelve antiquated iron bookstands which were then erected down the centre of the room, so the Hope Library once more became a reality. Carpenter also initiated plans for a new building to accommodate the Department and provide for the growing needs of entomology. This did not materialize, although for many years it was regarded as a viable proposition.

By means of a grant from the Christopher Welch Trustees, the Department purchased its first microscope, although Westwood did have a small microscope given to him by his schoolmaster, J. H. Abraham, FRS, which he put to daily use. Until then the pocket lens reigned supreme, and all that one needed otherwise to be an entomologist was a pair of pinning forceps and a packet of pins! The policy with regard to string and paper was augmented by economy in the use of envelopes. Although it cannot actually be proved, it has always been the contention that the Department was the originator of the economy label, so widely used during the war!

But alas, this delightful period in the history of the Department was brought to an abrupt close by the outbreak of the Second World War. The first consideration was the Types, which were all taken out of the Collection, packed into boxes, and moved to safety in a brick-vaulted chamber beneath the Museum. After some months it was discovered that this chamber was one of the main junctions of the central-heating system, and although it kept the specimens in admirable condition, the slightest damage to the pipes would have been disastrous for the Collections! Thus they were moved to the Ashmolean Museum where they spent the rest of the war half-way up a high wall on a small gallery—a most vulnerable position. All the valuable books were packed into boxes and stored in the Bodleian cellar.

Gradually the Department began to change. It still attracted specialists, but they mainly came for short visits only and when they left would marvel at the wealth of material they had discovered in the Collections or lurking in store-boxes yet to be dealt with.

In 1948 George Copley Varley became Hope Professor. Owing to the previous emphasis on mimicry, particularly in Lepidoptera, the exotic collections of that order had tended to monopolize the departmental activities, whilst the taxonomic collections of British insects (apart from Lepidoptera) had been much neglected. Professor Varley realized, because he needed reliable identifications of species of several orders involved in his ecological work, that the situation required correction. He therefore directed the work of the Department particularly towards the up-to-date rearrangement of the existing British collections, with the formation of separate reference and general collections, and the augmenting of orders which had been relatively neglected. The results have been far-reaching; many visitors have come to study these collections and some have used them in important revisions, including the Royal Entomological Society's *Handbooks*. Another project was the rehousing of the Type specimens, and in 1956 work commenced on the rehousing and cataloguing of the immensely important Pickard–Cambridge collection of spiders, a scheme for the catalogue being devised with the aid of advice from George Hazelwood Locket.

In 1960, public displays in the gallery being almost non-existent, the wall-cases were modified, and systematic exhibits covering all orders of insects were gradually installed. Other cases were also put on display to cater for such specialized topics as household and garden pests and insects of medical importance. A further activity was the reorganization of the disposition of the Collections and a rationalization of the chaotic mass of cabinets and store-boxes in the Wilberforce Room (which bore some resemblance to Manhattan skyscrapers!).

Much amalgamation and rearrangement of the exotic and British collections took place during the tenure of Professors Poulton, Carpenter, and

Varley. The work was carried out by various members of staff, both academic and technical, supplemented by visitors and research students. Details of their work may be gathered from the Departmental Reports and the Reports of the Committee for the Scientific Collections in the University Museum. With the incorporation of the academic side of the former Hope Department with the Department of Zoology, and the retirement of Professor Varley in 1978, a particular era in this history came to an end.

Type material in the Collections

One of the most important features of the Collections, which has contributed to their international reputation, is that they contain a large number of 'Type specimens' (the original specimens upon which the scientific descriptions of species have been based). Such Types are, in the broad sense, the property of science, hence institutions which hold them are both privileged and bear a responsibility. In this country the Hope Entomological Collections house more Types than any other institution, with the exception of the British Museum (Natural History). It has been the policy of the Department to provide every possible facility for those wishing to study Types. They are regularly loaned to research workers all over the world, and the number of loans of material from both the insect and spider collections has increased annually. Many of the Types represent species described at an early date by Hope, Westwood, and their contemporaries.

As regards cataloguing of the Types, a start was made only in 1903, on the Coleoptera. Little was then done until Professor Varley initiated the preparation of a card index of the insect Types, and another for the very large collection of spiders. Work on these indexes has continued steadily. They have been of immense value (indeed indispensable) when dealing with the many requests regarding Type material. In the old days some Types were believed lost, simply because they could not be located. The total number of species of insects and spiders represented by primary Type specimens (that is, holotypes or lectotypes) is currently estimated at 14 000.

NOTES AND REFERENCES

1. William Sweetland Dallas (1824–90)—a systematic entomologist temporarily employed for a time at the British Museum preparing a catalogue of the Hemiptera which appeared in 1851 and 1852. He was at one time Curator of the Yorkshire Philosophical Society and then Assistant Secretary to the Geological Society of London.
2. William Edward Shuckard (1803–68), Librarian to the Royal Society and a member of the Entomological Society Publications Committee.
3. *Oxford Magazine*, 22 (1904), p. 409.

5. Hope Professors—1861 to the present day

John Obadiah Westwood

JOHN OBADIAH WESTWOOD (Plate 6) was born in Sheffield on 22 December 1805, the son of John Westwood (1774–1850), a medallist and die-sinker and an expert in engraving and the embossing of wood, ivory, paper, and cardboard, among other materials. His family having connections with the Quakers, he commenced his education at a Friends' School in Sheffield. In 1819 the family moved to Lichfield, where he attended the grammar school, and later to Chelsea. Initially he was apprenticed as an engraver, when he acquired his skill in accurate delineations. But in the autumn of 1821 he was sent to London, and at the age of 16 was articled to Walton's, a firm of solicitors who soon offered him a partnership, which he accepted. By November 1828 he was enrolled as Attorney of His Majesty's Court of King's Bench at Westminster, in mid November of that year he was admitted as Attorney in Common Pleas, and later that month he became a solicitor in His Majesty's High Court of Chancery. He also held a certificate to practise as an attorney, solicitor, proctor, or notary public. Up to 1831 he practised successively at the following three addresses: Jubilee Place, Chelsea; with Jackson Walton and Thomas Andros at 8 Warnford Court, Throgmorton Street, in the City; and at 5 Brunswick Place, Hammersmith.

However, his real interests lay elsewhere and he admits that he read more natural history works than law books; having some private means, which he augmented by writing and his artistic talent, he devoted himself to the study of entomology, crustacea, palaeography, archaeology, heraldry, and mediaeval art. Indeed, he became one of the greatest authorities on Anglo-Saxon and medieval manuscripts and art and excelled in reproducing old manuscripts and illuminations. Due to his considerable knowledge of ivories and inscribed stones, he was employed to catalogue the ivories at the South Kensington Museum (now the Victoria and Albert Museum). His works on these subjects include: *Palaeographia sacra pictoria* (1843–5); *Facsimiles of the miniatures and ornaments of Anglo-Saxon and Irish MSS* (1846); *Illuminated Illustrations of the Bible* (1846); *A descriptive catalogue of the fictile ivories in the South Kensington Museum* (1876); and *Lapidarum Walliae* (1876–9).

From the time the Cambrian Archaeological Association was founded in 1846 until Westwood's death in 1893 hardly a volume of the *Archaeologia Cambrensis* appeared without one or more contributions from him.[1] As an artist he was completely self-taught and developed the ability to portray such diverse things as an insect or rare illumination. It can perhaps be said of him that there never existed nor maybe will ever exist again one who combined so great a fund of general information with such specialized knowledge of numerous groups ('the last entomological polyhistor', as Lindroth stated in *History of entomology*, R. F. Smith *et al.*, 1973, p. 121).

Westwood's apparent connection with the Swann family—John and James Swann, paper manufacturers who owned mills at Wolvercote, Eynsham, and Sandford-on-Thames in Oxfordshire—and the nature of his father's business would explain some of the contents of the Hope Library Archives. These contain envelopes addressed to Mr Westwood as 'Artist and General Manufacturer' and 'Manufacturer of Embossed Paper', samples of embossed notepaper, and cards embossed with the head of George IV and his consort Caroline, some annotated with the price.

The notepaper, with its elegant designs, sometimes has various inscriptions such as 'ADORATION' and 'SINCERETÉ', 'OH LISTEN TO THE VOICE OF LOVE'; one delightful sheet has a flying cupid blowing a trumpet with the words 'L'AMOUR' 'LA VERITÉ' underneath with another pair of cupids. Another sheet bears the curious inscription 'CATHOLIC EMANCIPATION 1829'. Westwood's particular form of doodling was to draw a cupid in flight, complete with trumpet, and this appears on some of his manuscripts and books (Plate 7). One wonders whether a beautifully embossed notepaper with the name 'Eliza' at the top was specially made for his wife. In 1839 he married Miss Eliza Richardson, which prompted his remarks in a letter to Peter Ryland, 17 September 1839: 'I beg to inform you that a little domestic occurrence, succeeded by a tour through Germany & Switzerland has alone prevented the regular appearance of the forthcoming nos. of my work'. Eliza Westwood accompanied him on all his archaeological tours and assisted in making sketches and rubbings of inscribed stones to illustrate his *Lapidarum Walliae*. She died in 1882. In 1887 Professor Westwood presented her twenty-one-drawer collection of minerals and fossils to the Museum.

Westwood's earliest record (in his diary) of collecting insects was in 1820 and 1821 at Lichfield 'but did not know how to preserve them till I saw Dr. Wright's'.[2] In April 1821 Mr Buckeridge gave him one of the late Dr Wright's drawers of foreign insects, to which Westwood added his uncle's specimens from Demerara. He had obviously been brought up in an atmosphere where natural history held an important place. At the sale of Dr Wright's collection in 1821 he purchased insects, and in his diary he records 'LONDON. Oct. Having bought a beautiful foreign bird at Dr. Wright's sale exchanged it with Heslop for a few English Insects—*a complete take in*'. He continued painting, buying cabinets and insects, and collecting assiduously, visiting in particular Hackney Marsh, Battersea Fields, and Coombe (Surrey). He, A. H. Haworth and other well-known entomologists of the day, would meet to discuss, name, and exchange specimens. By 1825 there were regular monthly meetings and in between it was the custom to drink tea at friends' houses and discuss entomological topics.

Westwood's diary records his first meeting with Hope in March 1824 when the former was taking tea with J. F. Stephens, Hope and G. R. Gray also being present. Eventually, they became close friends and in August 1834 he entered into an agreement with Hope to 'attend & arrange his insects except Coleoptera one day per week at £30 per annum'; this led to his eventual appointment as Conservator of the Hope Collections and the first Hope Professor.

In June 1823 Westwood called on A. H. Haworth 'when he showed me his Coleoptera and prom'd to give me some'. In July Haworth 'called to see my Ins[ts] when he picked out some which he had not got & promised me some in exchange for them'. By late July Haworth had fulfilled his promise and provided him with a great many insects. In September Haworth gave him leave to draw many of the insects in his collection and by October Westwood had completed a manuscript catalogue of the Coleoptera. When Haworth's collection was sold in 1834 Westwood purchased many specimens, some of which he gave to Hope.

Westwood met his fellow entomologist John Curtis for the first time in December 1823, when he showed him his drawings. Gradually Westwood's reputation as an artist grew, and he prepared illustrations and made lithographs for many of the books now regarded as classics of that age. In 1826, 22 August to 14 September, Westwood recorded in his diary a visit to Oxfordshire, when he visited Mr Swann and spent part of the time collecting specimens. He returned to Oxford in June 1832 for the meeting of the British Association for Promotion of Science and read a memoir on the respiratory organs of Crustacea and 'collected a few insects at Ensham [Eynsham] Shotover Hill & Blenheim'. In 1838 he visited 'Wychwode Forest'. These are but a few of the many visits he eventually made when the Hope Collections were transferred to Oxford from London.

By 1839 Westwood aspired to a position at the British Museum to succeed George Samouelle[3] and Hope supplied a 'handsome' reference. Unfortunately he did not have the opportunity to use it as Mr Samouelle's suspension from duty through contravening certain directions was considered sufficient punishment by the Museum authorities and he was eventually allowed to return. Their loss proved to be Oxford's gain. Westwood was acquainted with Samouelle from 1823 onwards and at one stage was colouring copies of *The Entomologist's useful compendium* (1824). His association with the British Museum started around 1829 when he received his 'reader's ticket', and by October 1830 he was frequenting the collections and making sketches for his great work *An introduction to the modern classification of insects*, having commenced the manuscript in August that year. A few years later (1836) he was actively engaged preparing dissections for the Museum.

In May 1827 Westwood was elected a Fellow of the Linnean Society. He was also associated with the Entomological Society of London from its foundation in 1833 and in 1834 was elected Honorary Secretary, a post he held until 1847 when he resigned because the new by-laws required the Secretary to devote more time than he was able to give. He was President three times, in 1851–2, 1872–3, and 1876–7, and at the Jubilee Meeting in 1883 was unanimously elected Life President. He also acted as Honorary Curator from 1836 to 1843, served on the Council for many years, and was on the Publications Committee and Library and Cabinet Committee.

In addition to contributing numerous papers to the Society's publications, Westwood prepared many of the original drawings for works published by the Society, for which he received a fee. He also donated many drawings to the Society, a notable one being the illuminated drawing of the vellum title of the *Signature book*, a handsome quarto volume bound in red morocco. This depicts 'Strength, beauty and utility' represented by a stag beetle, a butterfly, and a bee, together with a drawing of *Stylops kirbii* (a parasitic insect of the order Strepsiptera), all these forming a surround to the signatures of the Duchess of Kent, and Princess (later Queen) Victoria. In the left-hand corner are the initials 'J.O.W.', and in the right-hand corner '1835', the date on which both the painting and signatures were added. The design for the Society's seal, *Stylops kirbii*, was drawn by Westwood according to a note in his diary for 5 November 1834. The first (original) Entomological Society (Aurelian Society 1745–1806) had selected the bee as its emblem, but as the Society had become inert 'we may very entomologically infer that the Bee of the old Society had been Stylopized, and had at length given birth to the *Stylops Kirbii*' on the formation of the new Society ('Address on the recent progress and present state of entomology' by Westwood at the Anniversary Meeting of the Entomological Society 26 January 1835). The address ended as follows: 'This much, however, I may safely affirm, that although others may far exceed me in talent, I will yield to no one in point of zeal either towards our favourite science or towards our Society as its organ.'

Westwood was also elected Honorary Member of many scientific societies in Europe, as well as in Canada and the United States, and of local societies in Britain and Ireland. A splendid collection of 'Diplomata Westwoodiana' was presented by his niece Mrs Eva Whitmarsh (née Swann) in 1930 and is preserved in the Archives. In July 1833 he was made an Honorary Member of the London Vaccine Institution 'In Testimony of the high value they place on his liberal co-operation with them in the Philanthropic Cause of Vaccination'. His outstanding merits were also recognized by the Emperor of Brazil, Pedro II, who made him a Knight of the Imperial Order of the Rose, after visiting the Department in July 1874.

Westwood refused on several occasions to be nominated for election to the Royal Society, but in 1855 he was awarded their Gold Medal for his work, *An introduction to the modern classification of insects*, largely through Darwin's influence. However, he never sympathized with the theory of natural selection, and indeed on one occasion remonstrated with the authorities of Jesus College for allowing such a dangerous volume as the *Origin of species* to fall into the hands of a young undergraduate. The undergraduate in question was Edward Poulton, who was later to become one of the leading exponents of evolution and natural selection and to succeed Westwood as the second Hope Professor!

The accuracy of Westwood's work is beyond question and not many have made so few mistakes. There is, however, one incident which might have discomfited a more sensitive and less respected man—the 'Gateshead Flea'. This insect, which had been found dead in a bed at Gateshead, he had pronounced to be a gigantic flea and had described it as *Pulex imperator*.[4] Subsequently, however, at the Entomological Society of London,[5] he reported that he had examined the insect more minutely and had ascertained that it was a very young *Blatta* nymph, much distorted by being crushed flat in a rather oblique position, and with most of the limbs broken off. A small portion of the base of one of the multiannular antennae appeared to be a part of the mouth, but on microscopic examination of this and portions of the legs still remaining, it became evident that the insect was *not* a flea, and on dissecting the mouth, its true character was at once detected!

During the early years of his association with Hope, Westwood resided at Chelsea and then Hammersmith, but when the Hope Collections were removed to Oxford he became a frequent visitor and kept an eye on them before being officially appointed Conservator, and subsequently Hope Professor of Zoology. On taking up residence in Oxford he received

scant welcome from the Dons; the exclusiveness of that time being further aggravated by his Nonconformist origin and opinions, until rebuked by Richard Michell the Public Orator who reminded his friends that their new colleague was 'not sectarian but insectarian'. The good-humoured simplicity of his manner and his unfailing amiability to all who sought enlightenment in his department soon won mens' hearts and he became as popular as he deserved to be.[6]

His enthusiasm encouraged many to take up entomology—as a result of a meeting with Westwood, Henry Edwards of New York wrote that it 'Spurred me on to the study of entomology & thus gave me the greatest happiness of my life'.

Westwood took an active part in University affairs. His name was placed on the books of Magdalen College in 1858 and he was elected an Honorary Fellow in 1880. He was awarded an Honorary MA by the University in 1858, and the degree of MA was conferred on him by Decree of Convocation in 1861. He was particularly interested in building up the collections in the University Museum and donations of all disciplines came to him from far and wide. It is perhaps not generally known that he was largely instrumental in obtaining the presentation of the Ethnological Collection of General Pitt-Rivers to the University in 1882.[7] After Westwood's death his collection of shoes, including 'pampooties' from the Aran Islands and seventeenth- and eighteenth-century English ladies' shoes, was purchased for the Department of Ethnology.

Westwood also inaugurated a scheme whereby he made up collections of insects from duplicate material for use in schools—the first was a collection of British Coleoptera which he sent to the college at Clifton, Bristol. He was President of the Oxford University Entomological Society from 1859 until 1872, when the last unconfirmed minutes appear on 2 May: 'After some conversation about the prospects of the Society and summer excursions, the meeting adjourned', and apparently the Society did not meet again until 1922 when it was revived.

Publications

Westwood's writings on entomology began as early as 1827, many being published in the *Zoological Journal* and *Loudon's Magazine of Natural History* (several of the original drafts are preserved in the Hope Library Archives). He also contemplated publishing a new entomological journal. The draft of the prospectus is given in Plate 8, but the journal never materialized.

Apart from other interests, during the years 1830–2 Westwood was engaged on the entomological portion of a work entitled 'Encyclopaedia of Zoology', edited by William Swainson, to be published by Longman, Rees, Orme, Brown, and Green. An agreement was signed in March 1831 and Westwood was to be paid '10 guineas per sheet calculated from the largest type of Loudon's Encyclopaedia of Gardening'. From the outset, however, it was not a harmonious relationship. Westwood clearly did not endorse Swainson's devotion to MacLeay's theories, and felt that the space allotted to the insects was not commensurate with the huge size of the group. Swainson wrote many letters impressing on Westwood the necessity of conforming to his plan, over which they obviously did not see eye to eye. As a result Longman's paid Westwood for the work already done and both the collaboration and the work were discontinued.

Westwood also contemplated a companion volume to his *An introduction to the modern classification of insects* to cover the Crustacea and Arachnida, if advantageous terms could be arranged with the publishers. He also conceived the idea of a 'Genera Insectorum', which is represented in the Archives by extensive manuscript notes, and profusely illustrated by drawings and paintings. Neither of these works, however, came to fruition.

His publications were so numerous that only a brief account of the more important items can be given here. The list in Horn and Schenkling, *Index Literaturae Entomologicae*, Serie I (1928), occupies nineteen pages, and in Derksen and Scheiding, ibid., Serie II (1963–1975), four and a half pages. Outstanding items are:

The entomologist's text-book (London, 1838).
An introduction to the modern classification of insects, 2 vols. (London,

1838–40). This was known in Germany as the 'Entomologist's Bible', so named by Dr Schaum. The work appeared in parts, some of which were accompanied by parts of Westwood's *Synopsis of the genera of British insects*. Sheet B of the *Synopsis* was published in May 1838 with no. 1 of the *Classification*, sheet C in July 1838 with no. 3, sheet D in November 1838 with no. 7, sheets E and F in June 1839 with no. 13, sheet G in January 1840 with no. 15, and sheets H, I, K, and L in June 1840 with no. 16. (Note by Westwood in his copy of the *Classification*).

Arcana Entomologica, or illustrations of new, rare, and interesting exotic insects, 2 vols. (London, 1841–5).

Thesaurus Entomologicus Oxoniensis ; or, illustrations of new, rare, and interesting insects, for the most part contained in the collections presented to the University of Oxford by the Rev. F. W. Hope (Oxford, 1874). It is not generally known that this work appeared in parts—Part I, pp. 1–56, plates 1–10 was published in December 1873; Part II, pp. 57–112, plates 11–20 in March 1874; Part III, pp. 113–68, plates 21–30 in June 1874; Part IV, pp. 169–205, [i]–xxiv, plates 31–40 in September 1874.

By March 1835 Westwood had 'Completed my part of Vol. 2 of Murrays Nat^l. Histy Ins^{ts} & sent it to Gray—To receive for it 25.0.0.' *The natural history of insects*, vols. i and ii, was published by John Murray (London, 1829–35 [?1838]) in the Family Library series, nos. VII and LI. A note in Westwood's handwriting in vol. ii states 'Half of this Volume by J. O. Westwood. The remainder by G. R. Gray. See Table of Contents.'

In the preface to vol. 1 (p. viii) of his *Arboretum et Fruticetum Britannicum* (1835–8) (which appeared in sixty-three parts and also as a set of eight volumes in 1838) J. C. Loudon states 'received the assistance . . . of J. O. Westwood, Esq., F.L.S., Secretary to the Entomological Society, for descriptions and drawings of the Insects infesting different species . . .'.

Other contributions by Westwood include:

Lithography of the plate for Ingpen's *Instructions for collecting, rearing, and preserving British insects* (1827). Articles in *The British Cyclopaedia* by Partington (1835–37) and drawings for Jardine's *Naturalist's Library* (1834–43).

Figures for Stephens' *Illustrations* (1827–46) (for which he was paid 6*s*. 8*d*. per figure), Hope's *Coleopterist's Manual* (1837–40), Bell's *Crustacea* (1853), and Wollaston's *Insecta Maderensia* (1854).

Descriptions of insects for J. F. Royle's *Illustrations of the . . . natural history of the Himalayan Mountains*, vol. i (1839).

'The introductory observations, notes *passim*, description of the plates & the drawings & engravings' for P. B. Brodie's *A History of the Fossil Insects* (1845).

Westwood also prepared new editions of the following works:

D. Drury (1770–82), *Illustrations of exotic entomology*, vols. 1–3 (1837).

E. Donovan (1798), *Natural history of the insects of China* (1838).

E. Donovan (1800–3), *Natural history of the insects of India* (1838. For the two Donovan works he was paid £22. 10s. plus a copy of each volume.)

Moses Harris (1794), *The Aurelian* (1840).

H. D. Richardson (1847), *The hive and the honeybee* (1852).

He also collaborated with:

E. Blyth *et al.*, *Cuvier's animal kingdom*: *The articulated animals* (1840 and 2nd edn. 1848).

H. N. Humphreys, *British butterflies and their transformations*, 2 vols. (1841) and *British moths and their transformations*, 2 vols. (1843–5).

E. Doubleday, *The genera of diurnal Lepidoptera*, 2 vols. (1846–52).

C. Spence Bate, *A history of the British sessile-eyed Crustacea*, 2 vols. (1863–8).

In addition, Westwood published many articles in the *Gardeners' Chronicle*, as well as answering numerous queries in his capacity as 'Entomological Referee', a position he held for almost fifty years.

Travel

Westwood travelled extensively in Europe, especially between 1830 and 1847, and was probably better known on the Continent than any contemporary Oxford man, with the exception perhaps of Dean Liddell, the father of Alice of *Alice in Wonderland*. Westwood spoke French and German fluently and was in touch with all the well-known naturalists of the day, such as Latreille, Cuvier, Audouin, and Guérin-Méneville. On his trips he would study important collections, make copious notes and sketches, and visit libraries in order to obtain information from obscure books and journals. He also acquired many important works for his own and Hope's library, as well as insects for their collections, and later on for the Oxford Museum. He visited Paris on numerous occasions. In the Archives there is a small pencil sketch of Latreille made while at the Institut de France in September 1830; he also visited the renowned Jardin des Plantes. In 1832 he returned to Paris with G. R. Gray, where they met Latreille, Cuvier, and Audouin.

On 1 October 1835 returning home from an extensive European tour, he wrote to Hope:

My route has been as follows. Hamburgh [*sic*], Kiel, return to Hamburgh [*sic*], Berlin, Potsdam, Leipzig, Frankfort on the Main (where I heard of your visit from Dr. Ruppell) Mayence, Bonn (where the Meeting of German Naturalists has been held that year & where I met Audouin, Heyden, Charpentier &ᶜ) Cologne, Aix la Chapelle, Brussels & Antwerp. You may be sure that I have not been at these celebrated places without making the best of my time—As far as Entomology is concerned I have every reason to be most highly gratified—I have seen the Fabrician Collections—from which I have a great mass of notes—of books I have brought home 5 large Parcels including DeGeer—Panzer, Stoll—Fabricius Wiedemann & an infinity of other, standard authors—& which I have by rummaging out the antiquary Booksellers managed to procure at about one sixth of the English selling price—Of Insects I have procured still more valuable materials—my notebook being filled with Sketches & descriptions of the unique Insects in the Berlin & other Collections & my boxes being loaded with types of many of the rarest Genera, especially in Hymenoptera Diptera Hemiptera &.c

In the Archives there is a memento of his visit to Kiel—a single sheet of notes at the bottom of which is written: 'Manuscript of Fabricius given to me at Kiel 27th August 1835 by one of his two sons Dr. Fabricius of that place residing near the Cemetery on the road to Hamburgh [*sic*] in which Fabricius was buried. There is no tomb to his memory. J. O. Westwood.'

In his Presidential Address to the Entomological Society of London for 1872 he reminisces about his visit to Brussels: 'Professor Wesmael . . . One of my pleasant entomological recollections is that of meeting him one day in Brussels starting off on one of his collecting excursions, attended by a brace of gigantic St. Bernard dogs as a defence against the wolves in the Belgian forest, to which he was bound.'

In July 1837 he returned once again to France during which he went on an entomological excursion in the woods around Sèvres with Audouin and Brullé.

In September 1862 he was in Milan, and while at the Ambrosian Library copying an illumination from a fourth-century manuscript of Homer he took a male specimen of the Bethylid wasp *Cephalonomia formiciformis* which he found crawling over the vellum—an event he communicated to the Entomological Society. (*Observations on the Hymenopterous genus Scleroderma, Klug, and some allied groups* (1881), p. 128).

Other places Westwood visited in 1869 were Copenhagen (to see the Westermann Collection), Stockholm (for the Boheman Collection), St Petersburg, and Moscow; in 1874 he saw the Herrich-Schäffer Collection in Ratisbon (Regensburg), and also visited Geneva.

Reminiscences about Westwood

One of Westwood's maxims was 'Waste not, want not'. He was parsimonious regarding stationery—envelopes were re-used and his manuscript notes and drawings were often made on the back of old prospectuses, circulars, Great Northern Railway Scrip Certificates, and even a few on the back of French Revolution *Assignats*!

Regarding nomenclature, Westwood always recommended a study of the principles laid down in *Philosophia Botanica* (Linnaeus, 1751) and *Philosophia Entomologica* (Fabricius, 1778). He also thought a Hindu god was as worthy of having an insect named after him as any of the Greek or Roman gods or goddesses.

The Reverend O. Pickard-Cambridge visited the Museum one summer's afternoon and was invited to attend the lecture the Professor was giving the following morning on insects injurious to gardens. This being a popular subject, a good audience was expected.

Ten minutes before the hour of the lecture . . . I found the Professor completing his arrangements, and making a final disposition of his beautiful drawings and specimens. We remained there chatting for some little time, but no students or other audience appeared. Half an hour passed; still no arrivals. But the Professor was hopeful (was he not *Hope* Professor? but such a horrible joke could not occur to *him*): 'They will come presently; they are often rather late.' A gentle knock is heard at the door at last. 'Come in'; but no one coming in, the Professor goes to the door. 'Is this Professor Westwood's lecture-room?' asks a little timid voice. 'Yes, ma'am; we are all waiting.' And the Professor returns, followed by a little, rather elderly, frightened-looking lady, who is duly placed in a front seat; whereupon, without moving a muscle of his countenance, the Professor begins, and goes through an excellent and interesting lecture, with this little old lady as his whole audience; for it was only by being employed in assisting him with his drawings and specimens that I could restrain myself from exploding at the absurdity of the whole thing. First and last the Professor was as serious as if the whole University were before him.[8]

Although Westwood was thought to be lacking in humour, this is not borne out by various annotations in his manuscripts. For example, on his brochure regarding beehives at the Great Exhibition of 1851 he has written 'Drones not allowed'. Incidentally, he was a keen beekeeper and hardly ever used conventional protection, merely smoking a cigar while attending to the hives. He had also been known to cut combs from a hive without even that protection. Although a sincere Christian, he was opposed to anything savouring of clericalism, and on being asked to what religious sect he belonged, replied 'Sir, I am an Insectarian'.

In public speaking Westwood sometimes made curious mistakes. At a Gaudy Dinner at Magdalen College, he remarked in a speech, 'Well, gentlemen, as we all know, man arranges, but God disarranges'.[9] E. B. Poulton recalled a meeting of the Ashmolean Society when the late Professor Westwood read a paper on the great pest of the vine, the *Phylloxera*, and spoke of it as a curse—'for I look on wine as a good gift of God which maketh glad the *face* of man', and his cheery countenance seemed to bear out the fact. But Professor Rolleston, who held extreme views on temperance, spoke at the same meeting and called *Phylloxera* 'that good gift of God, the Phylloxera'.[9]

Apparently Westwood was never able to pronounce the letter 'h'. Many years ago Dr E. Burstal, living in Oxford as a small boy, was taught how to set out and care for insects in Professor Westwood's department. In 1967 he very kindly sent this quotation from a lecture on fleas given by Westwood: 'If I could 'op as 'igh as a flea—in comparison with our 'eights I could 'op to the top of 'Eadington 'Ill in an 'op and an 'arf' (Plate 9). Plate 10 shows an interesting cartoon found among Westwood's personal papers. Plate 11 shows mementoes of the Westwood era.

In a letter to Edward Saunders dated 20 April 1884 Westwood recalls an incident concerning drawings prepared by S. S. Saunders for a paper he published in the first volume (1836) of the *Transactions of the Entomological Society of London*. These drawings

which I think were amongst those stolen from Bond Street & which (after it being broadly hinted that I was the thief) I had the pleasure to regain from the pawnbroker to whom the thief's wife had pawned them, having cut them up & mounted them into a scrap book—I restored them to the Society in whose possession I trust they remain. I tell you this history as it is worth being placed on record.

Declining years

With regret one reads of the declining years of Westwood's extraordinarily active life. In May 1884 he wrote to Edward Saunders and mentioned that he had fallen downstairs and smashed bones in his left elbow; he saw little chance of recovering the use of his arm 'in which case I fear I must bid adieu to any more serious Entom! work'. It was the first year in the entire history of the Entomological Society of London, with the exception of 1873, that nothing from Westwood's pen had been published by the Society. Although he recovered from this accident and tried to carry on as before, advancing years soon began to restrict his activities. Attendance at meetings in London and elsewhere became less frequent and by 1892 he had become so incapacitated he was only able to visit the Museum on a few occasions with the aid of a bathchair. In November that year Mr W. Hatchett Jackson, Radcliffe Librarian and Hope Curator, was appointed Deputy Hope Professor under Clause 9 of the Hope Professor's Statute.

Westwood died on 2 January 1893, and it is fitting to quote the following passage from the Fifth Annual Report of the Delegates of the University Museum:

The year 1892 was a calamitous one in this department. At its commencement Professor Westwood fell ill, and his illness unhappily proved a continuing one, which slowly gained upon him, and put a stop his active habit of daily attendance at the Museum. His lectures were completely suspended, and he became unable to superintend personally the business of the department. This forced inactivity was a source of great trouble to him, but he cherished hope to the end of his days, which came suddenly just as the year had closed, when Oxford lost the most widely-known member of the scientific professoriate, and the Hope Department its first head, the most distinguished entomologist of the century.

Westwood was buried in St Sepulchre's Cemetery, Jericho, Oxford. There is a plaque in St Andrew's Church at Sandford-on-Thames which states:

The Vestry was built in 1893 in memory of
J. O. Westwood, M.A.,
Hopeian Professor of Zoology,
By Eva and Edgar Dyke Whitmarsh.

A fine oil portrait of him, painted by William Rivière, was presented to the Department in 1876 by Mrs Hope and it now hangs in the Library.

Several well-known personalities aspired to succeed Westwood as Hope Professor. Among them was Charles Valentine Riley, entomologist with the United States Department of Agriculture, who had visited Oxford on several occasions and wrote to Westwood in December 1886, 'Remembering the lovely climate and surroundings of Oxford, I sometimes feel that I should like to take up your work there when you lay it down; for though I never could hope to cast such unequalled lustre on the position as you have done, I should do my best to be a worthy successor.' Westwood admired his work, and would have liked Riley to follow him, but the honour went to Edward Bagnall Poulton, whose love of insects had been fostered by his association with Westwood and the Hope Collections.

Edward Bagnall Poulton

Edward Bagnall Poulton, elected to the Hope Chair in June 1893, was already well known in Oxford circles through his association with Professor Westwood. He was born at Reading on 27 January 1856. Of his early education at two boarding schools he wrote: 'The memories are of punishments only and the broad principle that every natural inclination was wrong and dangerous', and he described the seven years spent at his third school as 'a long dreary interval in a happy life'.

At an early age he showed an interest in natural history and during the summer of 1873 came to Oxford three days a week to work in the University Museum. Poulton himself comments: 'Then, when working for a scholarship in 1873, I allowed myself the luxury of an hour once or twice a week with Professor Westwood and the Hope Collections and Library. The most illustrious entomologist of his day was extremely kind to me, and these hours . . . were a delightful treat.'[9] The same year he obtained a Natural Science Open Scholarship at Jesus College and graduated with first-class honours in 1876. He then held a Demonstratorship under Dr G. Rolleston, Linacre Professor of Anatomy and Physiology, until 1879, when he resigned on being awarded a Burdett-Coutts Scholarship in Geology. In 1880 he returned to zoology and held a lectureship at both Keble and Jesus College until 1889, when he resigned to devote more time to his own interests. Poulton soon became known as an ardent exponent of evolution and champion of Darwin's and Wallace's theory of natural selection. This he applied to a study of coloration in insects and eventually published an epoch-making textbook in 1890 entitled *The colours of animals*.

Poulton also had a reputation of being parsimonious over little things. If a carpenter came to do a small job in the Department, he would carefully count out the exact number of screws or nails needed and, if the carpenter lost any of these during the course of his work, he had to replace it with one of his own! Mr Green, an old Oxford cabinet-maker who used to do work for the Department from time to time, after one of these episodes solemnly swept up all the sawdust, carefully put it into a bag, and handed it to the Professor, saying with a perfectly straight face 'No doubt you will be requiring this, Sir'. String was another of his little idiosyncrasies; he would carefully save all the bits and pieces he came across, and they were all placed in a drawer in the Department. Harry Eltringham (who studied under Poulton), on one occasion having fruitlessly looked in the string drawer for a suitable piece, is reputed to have exclaimed 'Damn it all, Poulton, why don't you sell this lot and buy a ball?' Poulton never bought paper for notes, but utilized the backs of old forms and circulars for his manuscripts, pasting all sorts of odd bits together—a nightmare for compositors. In spite of these economies, however, he was a most generous man in many other ways.

His marriage to Emily Palmer in 1881 brought him great happiness. Their house at Oxford and holiday home in the Isle of Wight welcomed many friends. Sorrow came with the loss of four of his five children in the prime of life, but his thirteen grandchildren and the six great-grandchildren were a great comfort in his old age. Emily was the daughter of George Palmer of the Huntley and Palmer biscuit firm. Their biscuit tins (now collectors' items) are a feature of the Department—dozens of them are packed with unset specimens of Lepidoptera! On his marriage one of his contemporaries jocularly remarked 'he got the biscuit and the tin'.

Elected to a Fellowship at Jesus College in 1898, Poulton played a prominent part in college life. He was a generous benefactor—providing new hymnbooks for the Chapel, silver for High Table, and redecoration in 1928 of the Old Bursary, maintaining the original design. He so enjoyed college hospitality that he left a benefaction to perpetuate it, and the Poulton Dinner is held annually. A portrait of him painted by Rothenstein in 1927 also hangs in the college.

Poulton was an active member of leading societies and many honours were bestowed on him. In 1889 he was elected Fellow of the Royal Society, served as Vice-President 1909-10, and was awarded the Darwin Medal in 1914. Fellow of the Linnean Society 1887, and President 1912-16, he served on the Council from 1889 to 1917 and was awarded the Linnean Medal in 1922. In 1884 he became a Fellow of the Entomological Society of London, was President 1903-4 and 1925-6, and Honorary Life President in 1933. To this society he made many generous donations, including the writing desk which once belonged to Alfred Russel Wallace, and he was such a dominant figure at their meetings with exhibits and communications, that at one stage it was said the Society might just as well be called 'The

Entomological Society of Oxford'. Tributes have also been paid on various occasions by a great many foreign societies.

Poulton conducted a vast correspondence with all the well-known entomologists of the day. Amongst his many friends were E. Ray Lankester, Alfred Russel Wallace, Roland Trimen, Raphael Meldola, Henry Fairfield Osborn, and Viriamu Jones (first Vice-Chancellor of the University of Wales). This friendship inspired Poulton to write *John Viriamu Jones and other Oxford memories*.[9] R. C. L. Perkins was a pupil of his at Jesus College, and also a notable donor to the collections.

One particularly close friend was Professor James Mark Baldwin, Foreign Correspondent of the Institute of France, who to 'celebrate and consecrate our long friendship and . . . testify to my real affection for [you] and Mrs. Poulton' established in November 1920 the 'Edward Bagnall Poulton Fund for the extension and diffusion of knowledge concerning evolution, organic and social', which was 'to be administered by the Hope Professor of Zoology, at his discretion'. Baldwin paid £100 per year into this fund until November 1931, when he gave a final generous gift of £1000. Over the years this fund has been of inestimable help to both postgraduates and undergraduates undertaking expeditions to various parts of the world.

Although he published a vast number of papers, his activities occupied so much of his time that he was never able to produce the *magnum opus* which was expected of him, and which he obviously had in mind throughout his tenure of the Hope Chair. Whenever he published an article with plates he would order extra copies which would be carefully stored away ready for incorporation in the great work. Alas, these plates were never used, and most were eventually employed as scrap-paper by research students in the Department over many years. However, a few plates survive and are now preserved in the Hope Library.

On 1 January 1933 Poulton resigned as Hope Professor to devote his time to writing up the records on protective resemblance, warning coloration, and mimicry that had accumulated over many years. He was knighted in January 1935 in the New Year's Honours List for his contribution to science, and continued to publish articles and attend meetings of various societies right up to the time of his death on 20 November 1943. At one stage he was transported to and from college and the Museum by wheelchair and his experience and continued help in the Department proved invaluable. It is perhaps fitting to end this short biography with a passage from the obituary notice by A. E. W. Hazel which appeared in *The Jesus College Magazine* (December 1943): 'He was a man of old-world courtesy and of kindly heart who would forgive almost anything except disbelief in the doctrine of evolution', and an extract from the address given at the funeral by the Reverend L. B. Cross:

He was a great lover of nature: in it he found a deep religious experience. . . . He would never destroy even the commonest or the smallest of insects unnecessarily. If he had to kill, it was always in order to make a sacrificial offering, a sweet savour, on the altar of truth. . . . He was devoid of personal amibition, though, when honours came to him, he received them graciously, and with full appreciation. He played no part in politics, or in University administration. He was content to do his work as Head of a Department. He had a strong sense of academic and social obligation, and a deep appreciation of the traditions and privileges of University life.

Geoffrey Douglas Hale Carpenter

The third Hope Professor, Geoffrey Douglas Hale Carpenter, was elected to the Chair on Poulton's retirement in 1933.

Carpenter was born in 1882 at Eton College where his father, Dr P. H. Carpenter, D.Sc., FRS, was science master. Educated at the Dragon School and Bradfield, he eventually took his BA at St Catherine's, Oxford, in 1904. From Oxford he passed with a University Entrance Scholarship to St George's Hospital, London, qualifying MRCS and LRCP, and taking the degrees of BM and B.Ch. in 1908 and his DM in 1913, presenting as his dissertation some of the results of his study of the tsetse fly. After qualifying he held the usual appointments of house surgeon and house physician.

He entered the Colonial Medical Service in 1910, and was engaged by the Royal Society to undertake a study of the bionomics of the tsetse fly, which had within a few years caused 200 000 human deaths from sleeping sickness around the shores of Lake Victoria. In 1911 he took up residence in the Sesse Islands, in the north-west part of the lake, and the results of his investigations were published in the *Reports of the Sleeping Sickness Commission of the Royal Society* in 1912, 1913, and 1919.

In the First World War he served with the rank of Captain in the Uganda Medical Service and was awarded the MBE. On demobilization he resumed work on the tsetse fly, becoming a specialist officer for the control of sleeping sickness in Uganda.

Encouraged by his friend Poulton, while in Africa Carpenter had devoted all his spare time to the study of entomology, particularly the butterflies. He soon became interested in natural selection and especially mimicry, sending large collections for the Department and observations upon the habits of insects which Poulton communicated to the Entomological Society.

One of the most important discoveries made by Carpenter while he was in Africa was, in 1912, the proof by breeding that the forms of the butterfly *Pseudacraea eurytus* known as *tirikensis*, *hobleyi*, *terra*, and *obscura* are con-specific. This resulted in what was probably the first cable ever sent about a butterfly and bore the single word 'terra', this being the name of the form he had succeeded in rearing from a single egg which he had observed to be laid by a female form *obscura*. Carpenter wrote to Poulton that 'if this egg produces a *terra* you will have the proof you so ardently desire. . . . As of course there will be no time to write I will cable just the one word either *terra* or *hobleyi*.' It was *terra* and the famous cable was sent! This was the first step in proving by breeding experiments that these forms of *Pseudacraea eurytus* were all of the same species. On 19 August 1912, Poulton, writing to Carpenter, concludes: 'P.S. . . . After your cable I imitated your example & have just had a sleepless night! But it's only natural & to be expected after such exciting intelligence.'

His friendship with Poulton resulted in prolific correspondence between them over many years, and these letters are now preserved in the Archives. Occasionally Carpenter would break into verse and include a few humorous lines, such as the following in his letter of 9 August 1917:

> Oh Sesse Isles are fair to see
> > And so is Damba too
> But none of the patch can hold a match
> To the isle of the Bollaballoo!

and

> A Dudu went doodling along
> And he sang a peculiar song
> > 'I do as I do do
> > Because I'm a Dudu
> And not to do doodly is wrong'!

(*Dudu* is the Kiswahili for an insect.) But sometimes in a more serious vein he would write on important events, such as the sinking of the *Titanic* in 1912—'In memory of the Engine room and Stoke hold crew'—and on the occasion of the Armistice in 1918—'A message. From those that passed on to us who are left.'

On retiring from the Colonial Medical Service in 1930 he came to live in Oxford and visited the Department daily to work upon the vast collections he had sent home. He was in fact the natural successor to Poulton and was elected to the Hope Chair in January 1933 with effect from the first day of Trinity term. His inaugural lectures on 'Insects as material for study' were given in November 1933.

During his tenure of the Hope Chair, Carpenter also paid special attention to the role of birds as predators of butterflies and therefore shapers of mimicry, although McAtee and others strongly opposed the view that birds were significant factors. Carpenter collected an imposing volume of data proving that in many parts of the world visual hunters such as birds do prey heavily on edible species of butterflies but avoid those that mimic distasteful species. As a result of his researches he became acknowledged as a leading authority on mimicry, and he became intensely interested in the imprints of birds' beaks often found upon the wings of butterflies which had escaped from their attackers. And for a time this dominated the policy of the Department. He was much more interested in beak marks than he was in the butterflies themselves, and many meetings of the Royal Entomological Society were taken up with long communications and exhibits on the subject. At last the

Plate 1. Frederick William Hope. (Courtesy of Ashmolean Museum.)

Plate 2. Ellen Hope. (Courtesy of Ashmolean Museum.)

Plate 3. Carved fireplace in British Room. (Photograph by Derek Whiteley.)

Oxford, Easter Term, 1861.

THE HOPE PROFESSOR OF ZOOLOGY will deliver an Inaugural Lecture on Saturday next, the 27th April, at Two P. M., in the Geological Lecture Room of the University Museum.

TAYLOR INSTITUTION,
April 20, 1861.

Plate 4. Notice for Hope Professor's Inaugural Lecture, April 1861. (Photograph by Denise Blagden.)

OXFORD UNIVERSITY MUSEUM.

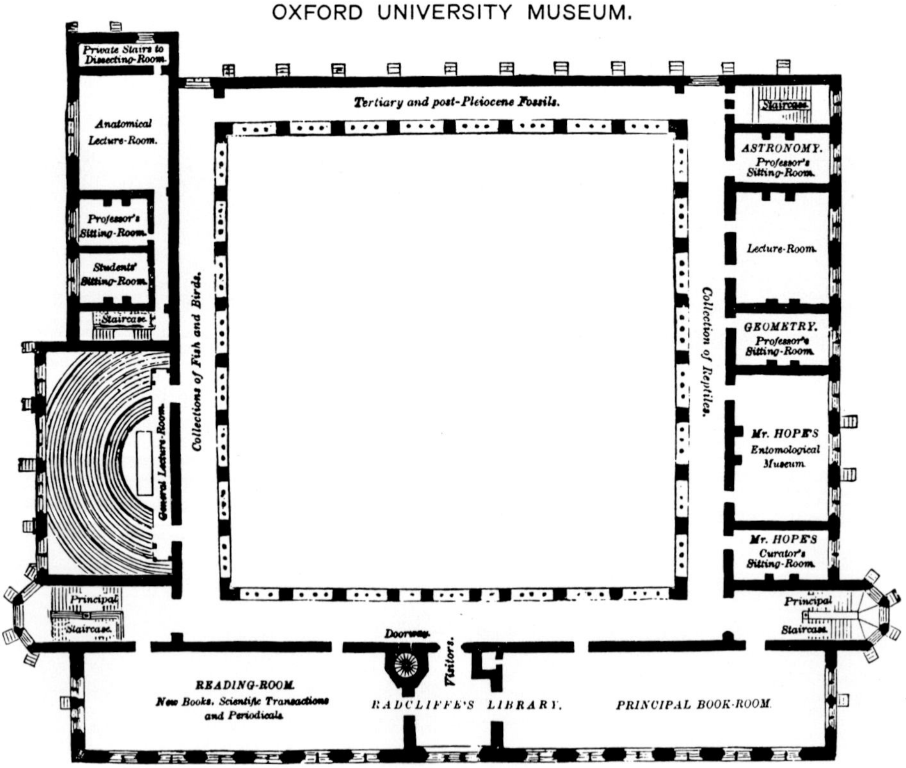

PLAN OF UPPER STORY.

Plate 5. *The Oxford Museum. The substance of a lecture by H. W. Acland*, 3rd edn., Oxford, 1866. (Photograph by Denise Blagden.)

Plate 6. John Oba-
diah   Westwood.
(Courtesy of Ash-
molean Museum.)

Plate 7. Samples of
Westwood's dood-
ling. (Photograph
by Denise Blag-
den.)

On the ~~(1st July 1827~~ (1st January 1828) will be published
(To be continued quarterly)
No. 1
Price (3/-?)
of
The Entomological Miscellany.
comprising
A popular account of the natural History
Properties Habits & scientific Characters of
The Families of Insects arranged according
to their Relations of Affinity
with Observations upon & descriptions of
new or rare Groups or individual Species
Notices of Entomological Publications
&c &c
interspersed with
Familiar Poetry original & selected
illustrative of the Subject
& embellished with
Coloured Figures of many rare & beautiful Species

By John Obadiah Westwood F.L.S.
Fellow of the Linnean Society of London
Member of the Zoological Club of the same Society
Corresponding Member of the Literary & Philosophical
Society of Sheffield &c

"Go to the Ant thou Sluggard, consider her Ways
"& be wise" Solomon
"He who enlarges his Curiosity after the Works of
"Nature demonstrably multiplies the Inlets to
"Happiness" Dr Johnson.

Plate 8. Westwood's prospectus for a new entomological journal. (Photograph by Denise Blagden.)

Plate 9. Quotation from a lecture on fleas: 'If I could 'op as 'igh as a flea—in comparison with our 'eights I could 'op to the top of 'Eadington 'Ill in an 'op and an 'arf.' (Drawing by Miss Frances Penrose.) (Photograph by Denise Blagden.)

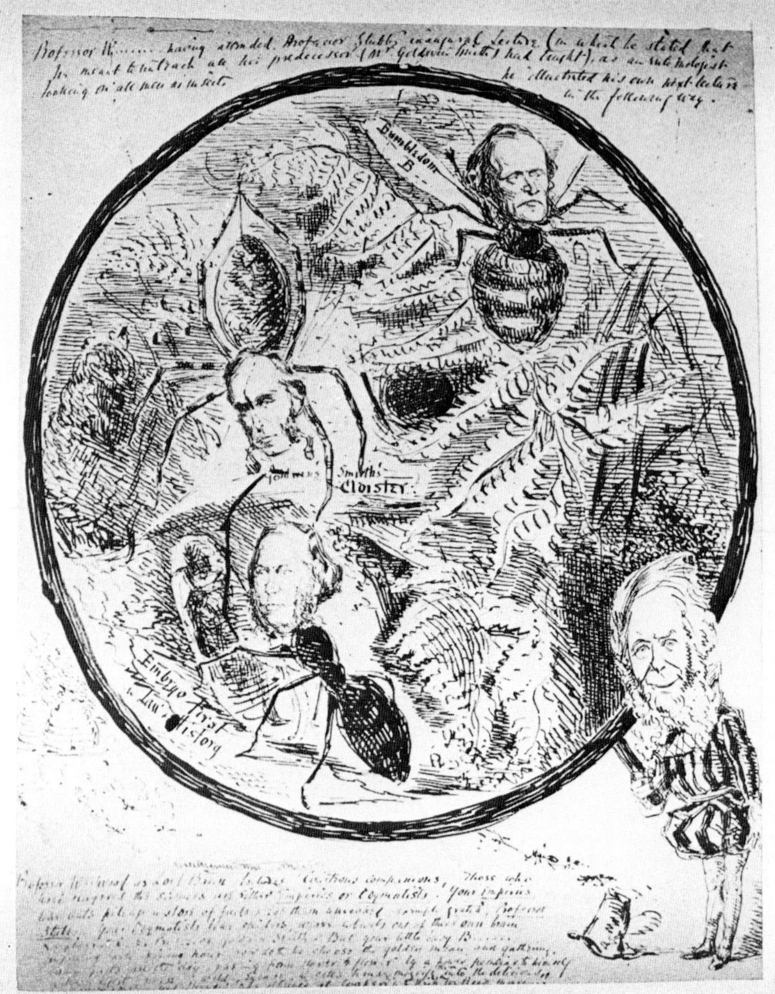

Plate 10. Cartoon found amongst J. O. Westwood's personal papers. The wording on the upper part of the cartoon is as follows:

Professor W... [Westwood] having attended Professor Stubbs' inaugural lecture (in which he stated that he meant to *unteach* all his predecessor (Mr. Goldwin Smith) had taught), as an entomologist looking on all men as insects he illustrated his own next lecture in the following way.

The wording on the lower part of the cartoon is as follows:

Professor Westwood as Lord Bacon lectures courteous companions, those who have fingered the sciences are either *Empirics* or *Dogmatists*. Your *Empirics* like ants pile up a store of facts and eat them uncooked, exempli gratia, Professor Stubbs. Your *Dogmatists* like spiders, weave cobwebs out of their own brain, exempli gratia, ex Professor Goldwin Smith. But your little busy B.... improving each shining hour now doth he choose the golden mean and gathering honey [spends] all the day, passing from flower to flower by a power peculiar to himself... [the rest illegible].

Lord Bacon is Francis Bacon, Baron Verulam, Viscount St Albans, 1561–1626, English philosopher, statesman, and essayist. The four figures in the cartoon are Goldwin Smith (top left), 1823–1910, 'Controversialist', Regius Professor of Modern History, 1858–66; Montague Burrows (top right), 1819–1905, Chichele Professor of Modern History, 1862–1905; William Stubbs (lower left), 1825–1901, Historian and Bishop of Chester and Oxford, Regius Professor of Modern History, 1866–84; and J. O. Westwood (lower right). (Photograph by Denise Blagden.)

Plate 11. Mementoes of the Westwood era. (Photograph by Derek Whiteley.)

Plate 12. View of Hope Library with the author and Dr M. W. R. de V. Graham, showing the Dillenius Cabinet and oil painting of Revd F. W. Hope. (Photograph by Derek Whiteley.)

Plate 13. Plate from *Archetypa studiaque patris Georgii Hoefnagelii*, 1592. (Photograph by Derek Whiteley.)

Plate 14. William John Burchell's painting of the interior of the wagon used during his South African journey, 1810–15. (Photograph by Derek Whiteley.)

Plate 15. Original painting of 'Telemachus Foem.', from *Jones' Icones*, Vol. III, LI. (Photograph by Derek Whiteley.)

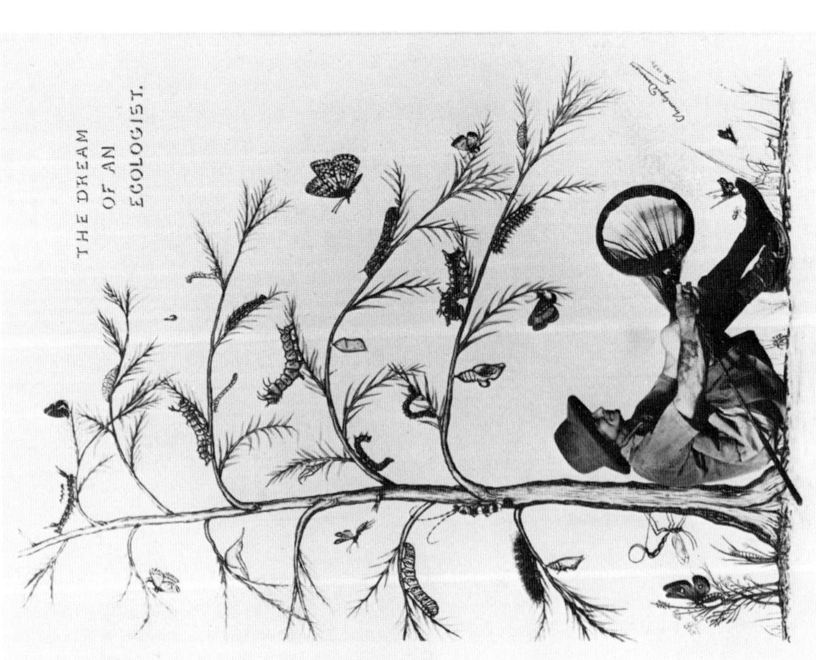

Plate 16. 'The dream of an ecologist.' (Drawn by Charles Dammers.) (Photograph by Denise Blagden.)

"Alas! — Those who in Museums contemplate the works of Nature do not think of all the perseverance, trouble and anxiety required before they are safely brought home"! —

H. Mouhot, three months before his death from jungle-fever on the River Mekon, territory of Laos, northwest of Cambodia.
Novbr. 10th. 1861. —

Col Tylden Curculionidae 16 Drawers

Plate 17. Note written by the Revd W. Tylden pasted on door of cabinet containing his Curculionidae. (Photograph by Derek Whiteley.)

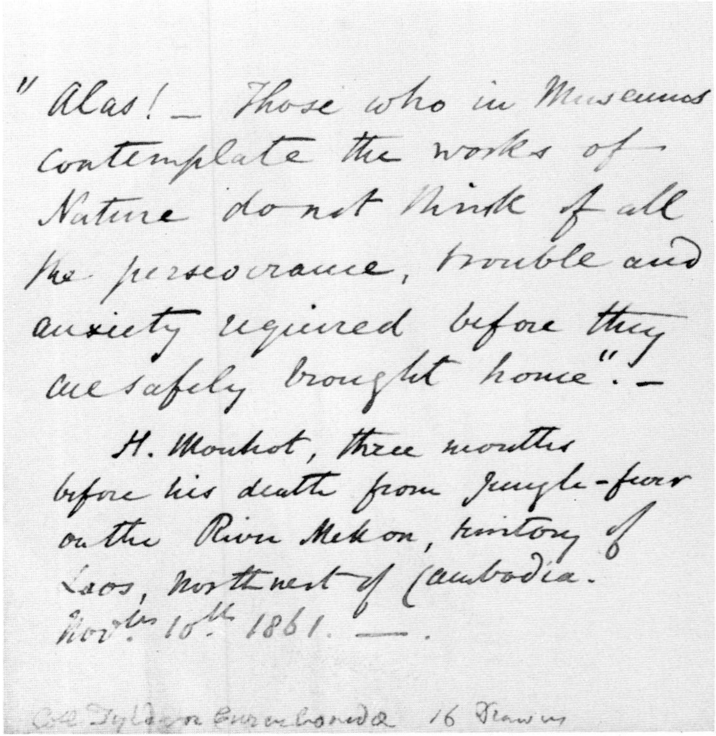

Fellows could stand it no longer and Carpenter was privately warned that unless he discontinued the exhibition of beak marks there would be a public protest against him. Far from being abashed by this, Professor Carpenter was overjoyed as it proved to him that he had produced overwhelming evidence of his case that birds *do* eat butterflies. In spite of these differences he was elected President of the Society for 1945–6.

During the Second World War Professor Carpenter was head air-raid warden for Cumnor. He also lectured to the troops on camouflage and tropical medicine and prepared booklets for use in the African campaign.

Throughout his life he published extensively, many of the articles being in the *Transactions and Proceedings of the Entomological Society of London*, and from his experiences in Africa he produced two books, *A naturalist on Lake Victoria* (1920) and *A naturalist in East Africa* (1925). After his retirement in 1948, Carpenter devoted the remaining years of his life to completing his last great work, *The genus Euploea* (*Lep. Danaidae*) *in Micronesia, Melanesia, Polynesia and Australia. A zoo-geographical study* (1953). He directed the preparation of the manuscript right through to galley-proof stage, mainly from a hospital bed, and fortunately he was able to read through the proofs before he died on 30 January 1953.

One of Carpenter's particular idiosyncrasies was that in the early spring of each year he would become extremely restless and each morning the first thing he said when entering the Department was 'Have you heard the chiff-chaff yet?'; he was never happy until that estimable bird had been heard, and even happier if he had been the first to hear it!

George Copley Varley

The chiff-chaff gave way to the cuckoo when George Copley Varley became Hope Professor on 1 October 1948. The latter was his 'call sign' and could be heard at all seasons of the year in Oxfordshire.

Varley was born 19 November 1910 and educated at Hulme and Manchester Grammar Schools. In 1929 he entered Sidney Sussex College, Cambridge, gaining first-class honours in Natural Science and Zoology, and his doctorate in 1938. From 1933 he held various positions in Cambridge—Superintendent of the Entomological Field Station, Research Fellow at Sidney Sussex College, University Demonstrator in Zoology, and Curator of Insects in the Museum of Zoology—and from 1940 to 1941 was engaged on the Wireworm Survey conducted by the Cambridge School of Agriculture.

During the Second World War he volunteered to work as civilian radio officer with the Army Operational Research Group of the Ministry of Supply and was mainly concerned with coastal centimetric radar. As a result of this experience he and David Lack (a contemporary Cambridge zoologist, who was later to become Director of the Edward Grey Institute of Field Ornithology at Oxford) produced a joint paper on the 'Detection of birds by radar'.[10] In

1945 he was appointed Reader in Entomology at King's College, Newcastle, a post he held until he came to Oxford.

As a result of his special interest in insect ecology, Professor Varley's research throughout his tenure of the Hope Chair was concerned with field and laboratory studies on the population dynamics of the winter moth and other defoliators of oaks, the knapweed gallfly, and the interrelationship between the various species of insect and their parasites dwelling in composite flower heads. The work was mainly centred on Wytham Wood, near Oxford, and he was assisted by the late George R. Gradwell (subsequently Lecturer in Forest Entomology, Oxford) and various research students working for their doctorates. The results of this work are embodied in many publications and especially the textbook *Insect population ecology*, by G. C. Varley, G. R. Gradwell, and M. P. Hassell, published in 1973. To cater for these research projects and assist with identifications, he initiated far-reaching changes in the organization of the British Collections, the formation of the general and reference collections being particularly valuable to all who wish to use the facilities.

Professor Varley travelled extensively and gave papers at meetings of the Ceylon Association for the Advancement of Science (1962), the International Biological Programme (Poland, 1966), and the International Union of Forest Research Organizations World Congress (Norway, 1976). It was during these travels that the extended range of the cuckoo came to be noticed!

Like Professor Carpenter, he too pressed for a new building to house the whole of Entomology. When this proved impossible he obtained various modifications to existing accommodation which resulted in the space being utilized more economically. At one time accommodation was envisaged in the new Zoology building, but this he would not take up as it would have meant leaving the Collections in the Museum, and thus divorced from the Library.

Professor Varley was an active member of the Royal Entomological Society of London and was President 1961–2. He was a keen supporter of their *Handbooks for the identification of British insects* and encouraged work by amateurs engaged on the project. The British Ecological Society also received his support and he was President from 1955 to 1957. On retirement from the Hope Chair in September 1978 he was elected Emeritus Fellow of Jesus College.

In addition to his busy working life, he was also a keen sportsman and apart from sailing and skiing, gliding was perhaps his favourite pastime. While in Cambridge he was a member of the University Gliding Club and after the war joined the Newcastle Gliding Club and gained his Silver 'C' Badge. In Oxford he managed to restore the prewar Gliding Club in 1951, and was President and Chief Flying Instructor for five years.

His interest in the Wytham Oak Survey did not end with retirement—he could be found most days in the Wilberforce Room working on the fine collection he amassed from Wytham Wood, building up information on the hosts which various parasites attack in order to understand the coexistence of so many different kinds of caterpillar on oak trees. Regrettably he died while this work was in progress—on 13 May 1983.

David Spencer Smith

The present Hope Professor, Dr David Spencer Smith, took up his appointment in February 1980. Although the research and teaching function of entomology has been integrated into the Department of Zoology, he remains in charge of the Hope Collection of Insecta and Arachnida, and of the natural history portion of the Hope Library.

Notes and References

1. Obituary in *Arch. Cambr.* (5) 10 (1893), p. 179.
2. Richard Wright, MD, surgeon and 'Proprietor of the Museum of Antiquities, Natural and Artificial Curiosities etc., Lichfield'.
3. George Samouelle (17??–1846), bookseller, who was on the staff of the British Museum *c*.1821–41.
4. *Proc. ent. Soc., Lond.* 4 May 1857, p. 70.
5. *Proc. ent. Soc., Lond.* 7 March 1859, p. 60.
6. W. Tuckwell, *Reminiscences of Oxford*, 2nd edn. (1907).
7. H. N. Moseley, letter, 26 March 1882.
8. *Entomologist*, 26 (1893), pp. 74–5.
9. E. B. Poulton, *John Viriamu Jones and other Oxford memories* (Longmans, Green, and Co., London, 1911).
10. *Nature*, 156 (1945), pp. 446–7.

6. Some personalities associated with the Department

Charles Swinhoe

By 1890 Poulton was prominent in entomological circles at Oxford, mainly through his friendship with Westwood over many years, and it was due to Poulton's initiative that Westwood allowed Colonel Charles Swinhoe to work on the exotic moths. Swinhoe, after an army career of some thirty years during which he accompanied Lord Roberts on his historic march to Kandahar, had retired and was living in Oxford. Between 1890 and 1898 Swinhoe devoted much of his time to arranging, cataloguing, and describing new species of exotic moths. Unfortunately he did not adhere to Westwood's instructions and this resulted in a sharp letter from the latter to Edward Poulton on 28 May 1892:

I made an agreement with Col. Swinhoe that I would do my best to secure an honorary degree of M.A. for him, when he had completed his task of the Eastern moths to my satisfaction. This has not yet been done & is still far from completion . . . his separation of the Eastern Collection into separate cabinets from the Africans &c contrary to my directions will give me a world of trouble to bring the whole into harmony, which existed as arranged by Saunders, before I bought it. . . . Your action in the matter is, in my mind, quite premature & your interference unsatisfactory under the circumstances.

However, Swinhoe's great two-volume work *Catalogue of the Eastern and Australian Lepidoptera Heterocera in the collection of the Oxford University Museum* appeared in 1892 and 1900 and Oxford duly conferred on him the degree of Honorary MA. Swinhoe also added to the collections many specimens of oriental Lepidoptera from the Khasia Hills in India and supplemented the salary paid to Westwood's assistant Shipp.

Amongst others associated with the Department at various times, the following are worthy of special mention.

Francis David Morice

The Reverend Francis David Morice (The Queen's College), a master at Rugby School, was in Oxford during the Christmas vacation 1893 and was asked by Poulton to assist in the revision and arrangement of the British bees. Poulton noted in his report that 'Fortunately the weather was then mild, and it was just possible to work, although with great discomfort, in the Hope Museum, where light is only available in positions in which the heat of the fire is imperceptible'. However, an 'excellent beginning was made' and he took an interest in the Collections right up to his death in 1926 when he bequeathed his collection of Palaearctic Hymenoptera to the Department.

Frederick Augustus Dixey

Frederick Augustus Dixey, MA, DM (Wadham College), also began his long association with the Hope Collections in 1893, paying particular attention to the Pierine butterflies, and was a regular visitor until his death in 1935. He published many memoirs as a result of his work in the Department, donated material to it, and was a Hope Curator from 1900 to 1932.

William Holland

William Holland was appointed as assistant in 1893. Originally a shoemaker at Reading, Holland was an example of the best type of working-man naturalist. On coming to Oxford he gave valuable assistance in the organization and arrangement of the great collection of insects of all orders. He turned his attention chiefly to Coleoptera, formed a collection of Carabidae which he presented to the Department, and was among the first to reveal the exceptional richness and interest of the beetle fauna of the Oxford district. His discovery of the first British specimen of *Gynandrophthalma affinis* in Wychwood Forest in 1899 was especially noteworthy. He retired in 1913 and died on 1 July 1930 at the age of 85.

Arthur Wallace Pickard-Cambridge

In 1896 Arthur Wallace Pickard-Cambridge, MA, D.Litt. (Balliol College), second son of Octavius Pickard-Cambridge the celebrated arachnologist, began to revise the British Coleoptera. He too was a good friend to the Department and in 1952 bequeathed to it a collection of historic importance made by his father and supplemented by himself. After the death of his wife in 1955, his trustees carried out his wishes and donated the sum of £1000 to the University of Oxford to keep the Arachnological Library bequeathed by his father up-to-date. This was accepted by decree in 1956.

Albert Harry Hamm

In 1897 Albert Harry Hamm, a printer by trade, was appointed by Poulton as an assistant. Originally a lepidopterist, on coming to Oxford he had become interested in the Hymenoptera and Diptera and formed vast collections, part of which he eventually presented to the Department. He paid great attention to the courtship of the predaceous flies of the family Empididae and his observations on their habits provided evidence of a complete evolutionary series. He was an expert photographer, for which he won several medals, was renowned for his prowess at bowls (being known as 'Tiger Hamm'), and was also a keen philatelist. Like Westwood, Hamm was also unable to sound the letter 'h'—this even spread to printing labels and there are some in the Hope Collections to this day bearing the name 'Ogley Bog' instead of 'Hogley'! In 1925 he was elected Associate of the Linnean Society and in 1942 Oxford University awarded him the honorary degree of MA. He retired in 1931 at the age of 70 and died on 9 January 1951 in his ninetieth year.

John William Yerbury

Colonel John William Yerbury came on the scene in 1898 and assisted with determinations and the arrangement of the Asilid flies. He also added to the Collections, the most notable gift being a series of British flies. His association with the Department came to an abrupt end in 1927 when he was killed in a motor accident.

Edward Saunders

Around 1898 Edward Saunders named many of the British Hymenoptera and helped to fill numerous gaps in the collection. His father, William Wilson Saunders, and his cousin, Sir Sydney Smith Saunders, also formed important collections, parts of which came to the Hope Department. Many convivial parties were held at his house at Reigate and on one occasion a guest stated 'I had a great honour last night . . . I was put to bed by Mr. Westwood'.

James John Walker

Commander James John Walker, having retired from the Navy, came to live in Oxford in 1904. From then on he was a constant visitor to the Department. During his naval career he travelled all over the world, collecting avidly whenever he could get ashore. His fondness for ships' cats earned him the nickname of 'Puss-olater'—he always said there were two things he could never resist: 'Sharpening a pencil and stroking a pussy cat'. He was best known as a coleopterist and in the field was noted for a trail of 'Walkerized' logs rent asunder in his search for wood-boring beetles. He spent much time arranging the collection of British Coleoptera and was always willing to assist with the general work of the Department. In August 1905 Oxford University conferred on him the degree of Honorary MA.

Walker was also a good botanist and maintained that he had tried almost every English wild berry poisonous or otherwise. One story he was fond of telling was how he began to draw a branch of mountain ash in fruit and when it was half finished he found that he had eaten nearly all the berries, one by one. Commander Walker was also renowned for his wit. Dr Harry Eltringham recalled the time they were looking over the English names of moths in the Hope Department, when they came across the 'Setaceous Hebrew Character (*Noctua c-nigrum*)' whereupon Walker, asked by his friend what such a character could be, instantly replied that the only 'Cetaceous Hebrew Character' known to him was Jonah! On another occasion, at a meeting of the Entomological Society of London when L. W. Newman exhibited specimens of a moth which had become so rare that he could only capture a few, and also some of the larvae which had all been parasitized, Walker, who was chairman, remarked that 'What with the Newmans and the Ichneumons the insect seemed to be having a bad time of it'!

In the Department he used to sit at the bench at the west end of the old British Room, humming to himself and breaking now and again into sea-shanties and home-made rhymes about his voyages, such as

> Just a song from 'Jerkins',
> When the breeze is low,
> And the poor old engines,
> Can't be made to go

and (slapping the back of his hand)

> Oh! Chrysops caecutiens,
> That's where our duty ends

a reminder of the painful attacks by clegs whilst collecting in the New Forest, and

> Oh flies oh flies
> You get in my eyes,
> Rifoldiritoodle-i-aye.

It was difficult to mention a part of the world with a coastline that he had not visited and he usually had some little anecdote to recount as a memento of his visit. He was on the editorial staff of the *Entomologist's Monthly Magazine* from 1904 and Editor-in-Chief from 1927 until his death on 12 January 1939 at the age of 88. Apart from various donations while working in the Department, he also bequeathed important material to the Collections.

Robert Walter Campbell Shelford

Robert Walter Campbell Shelford, MA (Emmanuel College, Cambridge), started work in the Department as Assistant Curator in 1905 and received remuneration through a grant from Magdalen College. He worked on the Orthoptera, particularly the Blattidae, also the vast collection of Bornean insects which he had presented during 1899–1901 while Curator of the Sarawak Museum, and assisted with the Library. In 1911 he gave up his post due to illness as a result of a fall, and died on 22 June 1912.

Joseph Joynson Collins

Joseph Joynson Collins of Warrington was appointed temporary assistant in 1905, the money being provided by Dr Longstaff. Collins had worked since the age of 13 as a wire-drawer in one of the local wire rope works and at this time was principally interested in the Lepidoptera. On arriving in Oxford he came under the influence of William Holland and Commander Walker and soon took up the Coleoptera. He amassed a collection of local insects which found a home in the Department. Collins's 'temporary' appointment proved to be very permanent! He retired at the age of 70 in November 1935 and died on 3 April 1942.

Roland Trimen, Charles Thomas Bingham, Frederick DuCane Godman, and George Blundell Longstaff

Several well-known personalities came on the scene during 1907. Roland Trimen assisted with the Ethiopian butterflies until 1909; Charles Thomas Bingham worked on the great collection of Ethiopian Hymenoptera until his death in 1908; Frederick DuCane Godman named all the American Hesperidae; and George Blundell Longstaff, MA, DM (New College), started work on the West Indian and American collections that he had donated. From 1904 to 1908 he personally defrayed the expense of an extra assistant for the Department to help with the ever-increasing congestion due to accessions, and in 1909 gave £2400 to the Trustees of Oxford University Endowment Fund towards maintenance and support of the Department. In 1912 he again visited Oxford and assisted for a while with the British Collections. Many specimens were donated during his lifetime, some of which he figured in published work. On his death in 1921 the Department was again enriched by a bequest of more material.

Harry Eltringham

Harry Eltringham, MA (Cantab.), D.Sc. (Oxon.), retired from business in 1908 and came to Oxford to study under Professor Poulton. He began research on mimicry in African Butterflies and worked in the Department for some twenty years in an honorary capacity. As a result of his work on the Collections he published extensively—*African mimetic butterflies* (1910), *A monograph of the African species of the genus Acraea* (1912), and many other systematic works on the Acraeinae and Heliconinae. He was also renowned for his histological work and superb artistic skill. Although he left Oxford in 1929 he maintained his link with the Department and was a Hope Curator from 1924 to 1937. Equipment from his laboratory was presented by his sister after his death in 1941. One other item of interest, preserved in the Hope Library, is the picture he took of Professor Poulton taken through the eye of a glow-worm.

Francis Cardew Woodforde

Francis Cardew Woodforde, MA (Exeter College), having retired from the headship of Market Drayton Grammar School in 1909, also came to live in Oxford. Professor Poulton asked him to assist with the arrangement of the collection of British Lepidoptera. This he readily agreed to do, devoting nearly all his time to the task and also adding to the collection until his death in

1928. Between 1920 and 1925 he published *Some notes on the collection of British Macro-Lepidoptera in the Hope Department of the Oxford University Museum.*

Edwin George Ross Waters

Professor Edwin George Ross Waters, MA (St Edmund Hall and Keble College), Professor of Romance Languages, was also a frequent visitor and willing helper, especially with the Micro-Lepidoptera. He died on 23 March 1930 at the early age of 39, and his fine collection of this group, including six volumes of mounted leaf-mines, was purchased by the Department for £100.

Karl Richard Hanitsch

Karl Richard Hanitsch came to Oxford after the First World War and gave freely of his time, helping in the Library and also working on the Orthoptera. Being an authority on the Malaysian Blattidae, his work and that previously done by Shelford made the Department a centre for the study of these insects. He built up and arranged the large collections and published many works. There being no adequate obituary because of the Second World War, it is fitting that a few details of this interesting man should now be given.

Hanitsch was born in Grossenstein, Germany, on 22 December 1860. Having passed his *Abiturium* at the Eisenberg Gymnasium he went to Jena to study mathematics, physics, and natural sciences and by 1885 had obtained the degree of Doctor of Philosophy and passed the State Examination in Natural Sciences (Upper Class). He came to England in 1886 to improve his prospects and held teaching posts in English schools for a short while. Unfortunately these posts were on 'mutual terms' (without renumeration) and he only received free board and lodgings in return for teaching. His finances during this period were very low, but in 1887 his aunt died and left him a small legacy, thus enabling him to redeem his microscope, which unfortunately he had been obliged to pawn earlier that year. He then applied for the post of Demonstrator at University College, Liverpool, and Professor Herdman engaged him at a salary of £80 per annum. His duties began in October 1887 and he was placed in charge of the Zoology Class. Early in 1895 he saw an advertisement in *Nature* for 'Curator and Librarian, Raffles Library and Museum, Singapore'. Having obtained excellent references from various people, including Professor Sollas at Oxford (they were both interested in sponges), he secured the position and sailed with his family for Singapore in May of that year. He became a naturalized citizen of the United Kingdom and Colonies in October 1910, and remained in Singapore until 1919 when he returned to England. Oxford University recognized his great contribution to science by awarding him an Honorary MA in March 1935. He continued working in the Department right up to his death on 11 August 1940.

Lawrence William Grensted

Professor Lawrence William Grensted, DD (Oriel College), Nolloth Professor of the Philosophy of the Christian Religion, became a regular visitor to the Department from 1932 until his death in 1964. He took a great interest in, and added to, the rather neglected smaller orders of insects. Indeed, the University Museum has much to thank him for—it was as a result of the 'Grensted Report' of 1952 that the Committee for the Scientific Collections in the University Museum came into being, whose function it is to

(a) exercise a general oversight over, and where necessary to supplement, the arrangements made for the care and maintenance of the collections . . .; (b) in consultation with the . . . professors to determine what part of the collections shall be exhibited to the public and supervise the arrangements for their display.

Bertram Maurice Hobby

It was in October 1934 that the first permanent post for an Assistant to the Professor and Librarian was approved by the Hebdomadal Council. This resulted in the appointment of Bertram Maurice Hobby, MA, D.Phil. (The Queen's College), to the position in 1935. Eventually he became Lecturer in Entomology and Librarian, a post he held until his retirement in September 1973. Hobby was no stranger to the Department. His research on predaceous insects and their insect prey commenced in 1929 under the direction of Professor Poulton, for which he gained the degree of Doctor of Philosophy in 1934. While working in the Department his interests ranged from the Asilidae to the bionomics of Empid flies, and during the war years he undertook research on behalf of the Government into the biology of Psocidae, pests of stored products. From 1937 he was on the editorial staff of the *Entomologist's Monthly Magazine* and as Editor-in-Chief took full control of the magazine from 1964 to 1981. As an editor he was meticulous and many amateur and professional entomologists have cause to be grateful for his constructive criticism of the papers submitted for publication. Dr Hobby was a Fellow of Wolfson College from 1965 to 1973 when he was elected to an Emeritus Fellowship. He died on 19 July 1983.

7. The Library

THE Hope Library (Plate 12) is an integral part of the Hope Entomological Collections because its nucleus was formed by Hope's gift of books from his unique personal library, later supplemented by Westwood's valuable collection of books and pamphlets, many of which had been purchased from the libraries of well-known naturalists such as J. V. Audouin and A. H. Haworth. Thus in the early stages of its history the Library, containing works dating from the sixteenth century, was one of the most important of its kind in existence. The enormous number of additions over the years by gift and purchase of both older works and modern textbooks and periodicals has made it still the most important entomological library in Great Britain next to that of the British Museum (Natural History) and the library of the Royal Entomological Society of London. Virtually every entomological classic is represented, often in a particularly interesting form, such as autographed copies, some in original parts as issued (for example Curtis's *British Entomology*), and many with manuscript annotations.

The Library is divided into three parts: books and periodicals which comprise approximately 13 000 bound volumes; catalogued pamphlets and offprints which number over 56 000 up to the present time; and the Archives. The Archives are of such importance that they are treated separately in Appendix A. The books are classified under various headings, including Natural History and Travel, Genetics, Evolution, Coloration, Ecology, General Zoology, Arthropoda, all the orders of Insecta, and Economic Entomology. The periodical section is especially noteworthy as it contains over 400 publications of which 137 are current serial titles; there are complete series of many of the more important entomological journals, and also some very rare older periodicals. Many of the current ones are not taken by any other library in Oxford.

There are also a number of non-entomological sections, such as Geology and Botany, which include books given by Hope, Westwood, and other donors, when interests were less specialized than today. The fine collection of pamphlets and offprints ranges from the late seventeenth century to the present time. The Library is therefore essential for curatorial and taxonomic research on the insect and spider collections because it contains not only a very comprehensive range of the older and rarer basic taxonomic works, but also a large number of most important manuscripts and letters which form a link between the Library and those collections.

Although the Library has been enriched over the years with works from the libraries of Pascoe, Rothney, Geldart, Morice, Meldola, J. J. Walker,

Hanitsch, and Poulton, to name but a few, special mention must be made of the bequest of the unique and extensive Pickard-Cambridge Arachnological Library in 1917, and the continuing donations, which commenced in 1981, from Dr L. G. Higgins of selected items from his extensive and valuable library. This latter magnificent acquisition further enhances the value of the Hope Library, particularly the Lepidoptera section, and is comparable with the original entomological libraries of Hope and Westwood. Practically all the works selected, such as Cyrillo, Poda, Clerck, Scopoli, Romanoff, and Rambur, were not represented in the Hope Library, or are copies containing significant variations from those already held. In accordance with his wishes they remained as a separate entity during his lifetime.

It is perhaps not widely known that the Hope Collections house a portion of another important library—the remainder of the Alfred Russel Wallace Library that was not acquired by the Linnean Society. It was donated by T. H. Riches in 1915 on the understanding that between the Linnean Society and Oxford the whole of Wallace's Library would be available to give an insight into the kind of man he was. The books cover a wide range of subjects, including a large number of works on spiritualism and the supernatural. The book most widely read by students of entomology was *Premature burial* but the collection as a whole has been used by members of the University Psychic Society.

Publications of particular note

Among the gems in the Library, the following works are of particular interest:

Archetypa studiaque patris Georgii Hoefnagelii (1592), modelled on the miniature paintings produced in the decorations of manuscripts illuminated for the Emperor Rudolf II (Plate 13). A manuscript note by Hope on the flyleaf states '. . . 47 Plates of Natural History beautifully coloured, with 26 of the original drawings . . . This Book belonged to the valuable library of Isaac D'Israeli Esqr. author of the Curiosities of Literature & many other popular works.' Another work from the Hoefnagel family is the *Diversae Insectorum Volatilium* by D. J. Hoefnagel (1630), comprising 16 black-and-white plates.

Naturgeschichte, Klassification und Nomenclatur der Insekten vom Bienen, Wespen und Ameisengeschlecht, by J. L. Christ (1791). This copy was purchased by Audouin from the library of Coquebert by whom the plates had been cut up and arranged in his 'Collection Iconographique' (see Appendix A, p. 71). It has marginal notes by Audouin and was eventually purchased by Westwood at the sale of his library in 1842.

Genera plantarum secundum ordines naturales disposita, by A. L. de Jussieu (1791). Our copy bears the following inscription on the inside cover: 'Hunc librum in itineribus Africa australi annis 1810 ad 1815 et in

Brasilia annis 1825 ad 1830 semper secum habuit. Gulielm. Johan. Burchell', and has pencilled annotations by him.

Klassifikazion und Beschreibung der Europäischen Zweiflügligen Insekten by J. W. Meigen (1804). This copy was originally presented by the author to Fabricius. It came into the possession of Latreille, and then Dejean, from whose sale it was purchased by Hope. Bound in the volume is a sheet of manuscript notes—those in pencil by Haworth and those in ink by Stephens.

Encyclopédie d'Histoire Naturelle . . . par le Dr. Chenu. Papillons (1853). This is Wallace's personal copy and was taken by him on his travels in the Malay Archipelago. It contains his signature, many drawings, and some manuscript notes.

Our copy of Ronalds's *Fly-fisher's entomology* is the 5th edn. (1856). Westwood has written in it 'The original specimens of insects from which the engravings in this book were made were given me by the author. J.O.W.'

There are many, many more interesting works, both entomological (such as the early books on bees and silkworms, including Malpighi's 1669 treatise *De Bombyce*), and non-entomological (such as Vallisnieri's 1715 *Istoria del Camaleonte Affricano*), but perhaps it is pertinent to mention a few other items preserved or housed in the Library, such as the small collection of daguerreotypes featuring well-known naturalists of the day; the busts of Linnaeus, Réaumur, Cuvier, Latreille, and Yarrell; the portraits of Linnaeus, Hope, Westwood, Burchell, Curtis, Kirby, Spence, Darwin, Wallace, and many other eminent people; the photomicrographs featuring the Ten Commandments, the Lord's Prayer, and Napoleon crossing the Alps; also four paintings of insects, flowers, and a lizard on 'rice paper' (not true paper but pith of *Fatsia papyrifera*) c.1850–90, from the Canton area of China, which were imported as curiosities via Hong Kong.

All these treasures make one realize that the Hope Library is in fact a miniature Bodleian and as such should be treasured and regarded as part of the national heritage. One can readily understand why Hope desperately wanted his books and collections to remain together. However, it is to be hoped that the Library, containing as it does such a wealth of books and periodicals both ancient and modern, and other historic memorabilia, will not be further split up by dispersing items to other departments for the sake of convenience, as has happened so often in the past.

Appendix A. Contents of the Archives

The archival holdings comprise a vast number of manuscript notes, drawings and paintings, and letters, many dating from the earlier part of the nineteenth century and some even from the eighteenth century. Only a selection of this material has been included and more detailed lists can be found in the records of the Hope Library.

The Hope–Westwood manuscripts were found originally tucked away at the back of old cupboards in the Department itself and in the corridor of the Museum. They consisted of masses of small bundles in a variety of wrappers, which were so dusty and fragmentary that unfortunately it was impossible to preserve them all. The bundles themselves, for the most part labelled by Westwood, have been kept intact. Amongst these manuscripts are letters from eminent naturalists, and the arrangement, as left by Westwood, has not been altered in any way. The main collection of correspondence has been collated, and early letters to Hope give some indication of his world-wide contacts when forming his 'Museum'.

Where archival material originating from an author has been incorporated with that of another, the fact is indicated at the end of the relevant entry— for example, (Westwood MSS), (Dale MSS), etc.

In 1979 it was discovered that some Westwood archival material had lain in the Smithsonian Institution in Washington for many years. Thanks to their kind co-operation this material has now been reunited with the main collection and fills several important gaps in the holdings.

In addition to the items enumerated in the following list, the archives contain thirteen boxes of plates dating from the seventeenth century onwards (many from well-known published works) and one box of original paintings acquired by Hope and Westwood. There is also a large collection of plates, some photographs, and paintings from the Poulton and Carpenter periods.

It should be noted that there is much important material in the Archives which appertains to the Collections and the two lists should be consulted in conjunction with each other.

ABBOT, C. 'Dr. Abbot's Entomological Calendar', 1798–1803 (extracted from his MSS by J. C. Dale) and notebook 'Lepidoptera Anglica cum Libellulis' (Dale MSS).

ACLAND, SIR H. W. Letters to Poulton, 1880–1907.

ADAMS, F. C. Lists of Diptera sent 1901–3 and typescript copy of 'Notes by the late F. C. Adams on some of the rarer Diptera found in the New Forest'.

ALBIN, — One sheet of paintings of insects 'F: Albin Fecit 1736' (Hope–Westwood MSS).

ANDREWES, H. E. Letters to Poulton, 1918–21.

Notes on Types in Chevrolat Collection, on Putzeys' Types of *Clivina* (Hope Collection), and on Types of eastern Carabidae in Oxford Museum (Hope Collection).

ANDREWES, H. L.   Letter to Carpenter, 1933, and catalogue of his Indian collection of moths.

ANDREWS, E. A.   A few fragments of notes from Poulton manuscripts.

ARNOLD, G.   Four sheets of original paintings of Chrysididae attributed to G. Arnold; also some correspondence.

AUDOUIN, J. V.   Letters to Hope, 1834, and Westwood, 1835; manuscript copy of 'Audouin. MS observations Entomologiques', copied by Westwood from original lent to him by Madame Audouin, 1852; catalogue of library; fragment of manuscript; fragment of letter from G. R. Gray to Audouin; letter from L. Dufour in offprint 36141 in Hope Library.

BACON, DR   *see* HOPE, REVD F. W.

BADCOCK, H. D.   Notebooks, drawings, and letters.

BAIRSTOW, S. D.   Letter to Westwood, 1882, also list of specimens sent from Cape of Good Hope; letter of 1891 concerning shipment of South African shells, and draft of Westwood's reply accepting collection and conditions.

BAKER, J. R.   Identification of insects from the New Hebrides by the Imperial Bureau of Entomology in 1923.

BALDWIN, J. MARK   Correspondence concerning E. B. Poulton Evolution Fund, *c.*1920.
   See also POULTON, E. B.

BARNETT, DR   List of insects collected in north-west Peru, April 1931.

BARRAS, J.   'These [3] Butterflies are pickled, they are not painted—that is to say the wings are natural—the bodies are painted', 1889.

BARROW, J.   Manuscript 'Cape of Good Hope Vol. 1. 1798'. Contents: 'A General Description of the Colony' and 'Accounts of the Revenues & Improvement thereof'. Note by W. J. Burchell: 'This MS Bought at Bohn's 28 Aug⸍ 1824 appears to have been written just previously to the publication in 180[1–4] of the First Edition of Barrow's Travels in Southern Africa. But subsequently in 1806 this MS has evidently supplied the materials (in many places, *verbatim*) for the *second* volume of the *second* edition of those Travels—5. 12. 43.' (Burchell MSS).

BATE, C. SPENCE   Manuscript notes and drawings by Westwood for *A history of the British sessile-eyed Crustacea*, 1863–8, by Bate and Westwood; also relevant letters (Westwood MSS).

BATES, H. W.   Letters to Westwood, 1864–7. Manuscript descriptions of Mantidae and notes on Oxford Collection, also some by Westwood; manuscript list of species of Gryllidae (Westwood MSS).
   Notes made by J. C. Dale on some species of *Libellula* sent from Para and Santarem, 1852 (Dale MSS).

BEAN, MR   For list of specimens in his collection, *see* BELL, T.

BEDDEK, MAJ. E.   Part of letter from R. C. L. Perkins to Poulton concerning specimens to be acquired for Hope Collections.

BELL, T.   Letter to Hope, 1839.
   Notebook on British Crustacea; catalogue of Crustacea in spirit; letter from J. S. Bowerbank to Bell, 1843, giving 'List of named specimens of Crustacea in Mr.

Bean's Collection'; letter from W. Brightwell to Bell, 1839, with original drawings; letters to Westwood (two 1861, two undated); letters from W. P. Cocks to Bell, 1845, 1847; 'Memoranda on Irish Crustacea'; 'Extract from Journal . . . 2nd May 1835' and note dated 12 May 1835 by R. Patterson; letters from H. Thompson of Belfast, 1843; sheets of printed numbers used by Bell; 'Memoranda on the physiological habits of stalk-eyed Crustaceans by Jonathan Couch, Polperro' (sent to R. G. Couch); miscellaneous drawings by R. G. Couch.

Catalogue of the Collection of Crustacea belonging to Professor Bell purchased and presented to the Hopeian Museum in 1862 by Westwood.

BERKELEY, REVD M. J.　Original drawings of fungi (also paper by him entitled 'Decades of fungi') (Hope–Westwood MSS).

BESSER, W.　Letters to Hope, 1828–33, including list of Coleoptera sent 30 June 1830.

BLAKE, J.　*see* MILLAR, H. M.

BLOMER, CAPT. C.　Entomological journal, 1820–35. Catalogue of Blomer's cabinet and sundry notes by J. C. Dale (Dale MSS).

In 1835 Dale purchased a cabinet containing 3215 specimens for £53. 11s. 8d., and in 1839 sold part of the collection to the Revd J. Streatfield for £35, who presented it to the Museum of the Margate Literary and Scientific Institution.

BLOOD B. N.　Notebooks, Mymaridae I, III–VI; Entomological Memoirs, 1929–30; original drawings and paintings; manuscript notes; miscellaneous photographs; two letters from A. H. Hamm; letters concerning purchase of collection.

BLUNT, E.　List of Lepidoptera in the cabinet of E. Blunt which later came into the possession of J. Curtis (Dale MSS).

BOND, F.　Original label of fossil crabs from Cambridge, 1864.

BOWRING, J. C.　Letter to Westwood, 1850.

Manuscript notes and drawings by Westwood for joint paper entitled *Notes on the habits of a Lepidopterous insect parasitic on Fulgora candelaria*, 1876.

BOYS, CAPT. W. J. E.　Letters to Hope, 1842.

Paintings and transfers of Indian insects and manuscript of paper on *Mutilla* read to Entomological Society of London, June 1843 (Westwood MSS).

BRIGGS, [J. H.]　Manuscript notes on breeding of *Liparis dispar* (Westwood MSS).

BRODIE, N. S.　Catalogue of part of his collection prepared by G. D. H. Carpenter and letters from Mrs Brodie, 1933.

BRODIE, P. B. and W. R.　Letters to Westwood (1853–4) concerning work on fossil insects (Westwood MSS).

BROWN, E. S.　Notebooks on Hemiptera (mainly aquatic); manuscript notes on East African Army Worm project and work in the Middle East; twelve diaries written in Madagascar, Seychelles, and Kenya.

BROWN, H. ROWLAND-　*see* ROWLAND-BROWN, H.

BROWN, S. C. S.　Typescript of notes on the Dale diaries, letters, library, and collection, copy of wills of J., J. C., and C. W. Dale, and other miscellaneous papers, presented by S. C. S. Brown, 1983 (Dale MSS).

BROWNE, REVD J. C.　Letter to Westwood, 1854, enclosing manuscript of 'Notes of English Papiliones found in India'.

BRUCK, E. VOM　List of beetles sent to Hope, and letter to Westwood, 1848, concerning gift of Coleoptera in return for kindness shown to him.

Brunner Von Wattenwyl, C.   For list of collection of Orthoptera from Rio Grande do Sul *see* Burr, M.; letter to Poulton, 1907.

Buckton, G. B.   Letter to Westwood, 1882; letters to Poulton, 1899, 1900, and from his daughter Jessie Buckton, 1906. Original paintings for his *Monograph of the Membracidae*, 1903, which contains descriptions of large numbers of specimens in the Hope Collections (Hope Library, bound in three volumes).

Bullock, G. H.   Letter to Poulton, 1921; list of butterflies collected on 1921–4 Mount Everest Expedition.

Buquet, L.   Lists of insects selected by Hope; also letters to Hope, 1834–5, and Westwood, 1850, 1868.

Burchell, W. J.   Letter to Hope, 1832; list 'Cetoines de Mr. Burchell' (?Audouin); letters to Westwood from J. D. Hooker and J. Phillips, 1865, concerning offer of Burchell Collection to Oxford.

   Burchell's personal copy of his book *Travels in the interior of Southern Africa*, 1822–4, with annotations (presented by J. P. Mansel Weale in 1897).

   Notebooks, letters, etc., covering his early years, stay on St Helena, and travels in South Africa and South America; nine sheets of pencil sketches mapping area where he travelled in South Africa; original paintings (Plate 14); legal documents. These manuscripts were presented in 1906 by his grandnephew Francis A. Burchell.

   Westwood's notes on Burchell Collection; transcript of St Helena journal and copies of Burchell's letters, with notes by Poulton and Mrs McKay; Poulton's notes on the Insect and Plant Collections, also on Brazilian insects; list of British insects with data from specimens; Poulton's manuscript notes concerning Burchell's honorary degree at Oxford; letters in connection with research on Burchell and his collections by Poulton and Mrs McKay.

Burney, Revd R.   Manuscript catalogue of moths in the cabinet of the Revd R. Burney as it was in November 1824 (Dale MSS).

Burr, M.   Correspondence, 1896–1953. List of collection and of Orthoptera received from Brunner von Wattenwyl.

Butcher, Mrs H. (née Burchell)   Notes by Burchell and Poulton on insects collected by her in Van Diemen's Land, and brought to England, 1836 (Burchell MSS).

Butler, Revd A. G.   Letter to Westwood, 1881; original paintings for *Descriptions of six new species of Diurnal Lepidoptera in the British Museum Collection*, 1865; letters to Poulton, 1897.

Calyo, S.   Drawings of fishes and Crustacea (also some plates) (Hope–Westwood MSS).

Cambridge, Sir A. W. Pickard-   *see* Pickard-Cambridge, Sir A. W.

Cambridge, F. O. Pickard-   *see* Pickard-Cambridge, F. O.

Cambridge, Revd O. Pickard-   *see* Pickard-Cambridge, Revd O.

Cameron, P.   Letters to Westwood, 1884, 1889, and lists of specimens sent by Cameron; plates for his *A monograph of the British Phytophagous Hymenoptera*, 1882–93; letters to Poulton, 1894, 1896.

Cantor, T.   Letters to Hope, 1841–2, 1846, and notes 'Extracted from Dr Cantor's Journal. Chusan Collections.'

Carmichael, Lord   Letter from Brunetti with list of material from Lord Carmichael, 1920.

Carpenter, G. D. H.   Notes, drawings, and photographs connected with his published and unpublished work whilst in East Africa and during his tenure of the Hope Chair; much correspondence, mainly on Lepidoptera and mimicry (including one box of letters from Carpenter to Poulton, 1910-30); 'Terra' telegram to Poulton, 1912 (see p. 49).

CARR, MISS ANNA   Paintings of animals, mainly invertebrates, and pencil sketches, 1835-7; presented by her greatnephew Major S. L. Norris, 1924.

CHABRIER FILS   Letter to Hope, 1835, with list of insects.

CHAMPION, G. C.   Diaries of captures, 1868-89; notes on localities visited in Colombia; original drawings of beetles; correspondence with Poulton, 1896-1918.
   Original drawings of spiders, correspondence, and information concerning *Biologia Centrali-Americana* (O. Pickard-Cambridge MSS).

CHAMPION, H. G.   Letters from his father G. C. Champion, 1916-25; a few other letters and miscellaneous notes.

CHAPMAN, T. A.   Letters to Westwood, 1868, 1870; drawings by Westwood including a few for *Some facts towards a life-history of Rhipiphorus paradoxus* by Chapman, 1870; letter from J. Hellins, 1870. *See also* Westwood's notes labelled 'Aptera', 1865.
   Correspondence with Poulton, 1890-1912.

CHEVROLAT, L. A. A.   Letters to Hope, 1822-7, 1838-42, and list of Coleoptera sent by Chevrolat, 1842; letters to Westwood, 1834, 1850-2.
   For notes on Types in Chevrolat Collection *see* ANDREWES, H. E.

CHILDREN, J. G.   Letters to Hope, 1834, 1837; *see* Westwood's notes labelled 'Aphaniptera' for sketch of flea in Mr Children's cabinet.

CHITTY, A. J.   Manuscript list of Chitty Collection by Claude Morley; letters to Poulton, 1906, 1907, 1921.

CLARK, BRACY   Letter to Westwood, 1842, and manuscript list of *Oestrus* by Westwood.

CLARK, E. J.   Manuscript notes and three notebooks on the ecology of British grasshoppers; miscellaneous letters concerning posthumous publication of his manuscript, 1945.

CLARK, HAMLET   Letters to Hope, 1848, and to Westwood, 1860-5; manuscript notes concerning Cryptocephalidae from New Holland in Hope and Westwood Collection.

CLARK, J.   Letter to Westwood, 1854, sending packet from Colonel Reid containing insects from China.

CLARK, J. W.   Letter to Westwood, 1869, concerning Mr Flower's Crustacea.

COFFIN, E. P.   Two notebooks '. . . on the habitats of his collection of Mexican insects . . .' (Westwood MSS).

COLLIN, J. E.   *see* VERRALL, G. H. and COLLIN, J. E.

COLLINS, J. J.   Notebooks and miscellaneous notes.

COMPTON, T.   'Passaliden-Club' Th. Compton del. W. A. Meijn lith.; original plate, issued separately, by Theodor Compton, illustrator of J. J. Kaup's *Monographie der Passaliden*, 1871, and a copy reproduced by René Oberthür at his printing works in Rennes, with omission of lithographer's name.

COOKE, BRIG. B. H. H.   Letters concerning bequest, 1944-7, and copy of agreement signed by Carpenter.

COQUEBERT, A. J.   'Collection Iconographique de Diptères et d'Hyménoptères recueillis par Mr Antoine Jean Coquebert' (in manuscript with cut-out figures from works by Fabricius, Réaumur, Schaeffer, Geoffroy, Panzer, Christ, Latreille, Coquebert, and Meigen, also some original figures drawn by Coquebert). This work was purchased by Audouin from Coquebert's son in 1837 and by Westwood at the sale of Audouin's library, 1842, for 25 francs.

COSTA, A. and O. G.   Letters to Hope, 1834, 1848–55, including list of insects offered to Hope; collation of *Fauna del Regno di Napoli* [1829–86] mainly by Westwood.

COUCH, J.   Three original paintings, 1859, 1867, and one no date (O. Pickard-Cambridge MSS).
'Memoranda on the physiological habits of stalk-eyed Crustaceans' (Bell MSS).

COUCH, R. G.   Miscellaneous drawings (Bell MSS).

COWAN, REVD W. D.   Letters to Westwood, 1882, concerning purchase of insects from Madagascar, also offer of birds and mammals.

COX, MRS E.   Letter to Poulton, 1896, concerning collection of flies and beetles from Tasmania.

CRAWFORD, F. S.   Letters to Westwood, 1882, 1888, 1889; notes, drawings and photographs of Australian Coccidae; also some correspondence with Miss E. A. Ormerod. Drawings of mites (Westwood MSS).

CRAWLEY, W. C.   Ant bibliography and notebooks.

CRÉMIÈRE, —   Letters to Hope and his desiderata list, 1842.

CROSSE, W. H.   Letters from E. M. Bowdler Sharpe to Poulton, 1902–3 and 1913, concerning his collection.

CURTIS, C. M. (Younger brother of J. Curtis).   Letter to Hope, 1836, concerning insects from Madeira. Original paintings of Coleoptera and larvae (Hope–Westwood MSS).

CURTIS, J.   Letter to Westwood, 1830. Economic and taxonomic manuscript notes with annotations by Westwood, includes notes used for *16 Reports on the insects injurious to agriculture*, 1841–57, and *British entomology*, 1824–40; original drawings and paintings; letter from G. Passerini, 1854; the notes etc. were purchased by Westwood from Mrs Curtis in 1863 (Westwood MSS).
Letters to J. C. Dale, 1819–62 (Dale MSS).
For Minutes of Third Meeting of the Norwich Entomological Society held at Mr Curtis's in 1810 *see* NORWICH ENTOMOLOGICAL SOCIETY.

DAHL, G.   Price list of insects, 1829.

DALBY, REVD A.   Letters to Carpenter, 1938, 1941, concerning collection and data for 1941 collection.

DALE, J. C. and C. W.   Unique collection of manuscripts covering virtually the whole of the nineteenth century and containing, apart from invaluable entomological information, many matters of sociological interest. The collection is so large that the following list has been restricted to the most important items:
Manuscript calendar, 1807–12.
Entomological diary, 1815.
Entomological diary, 1835–65.
Lists of localities entomologized by J. C. Dale, 1800–69.
Entomological diary 1860–72.
Entomological calendar of Dr C. Abbot, 1798–1803, extracted from his manuscript, continued by J. C. Dale, 1808–35.
Diary for years 1827–30—mainly travel.

List of correspondents.

Catalogue of J. C. Dale's cabinet of British insects.

Catalogue of moths in the cabinet of Revd R. Burney as it was in November 1824.

Revd C. Abbot's 'Lepidoptera Anglica cum Libellulis'.

Address book.

The late Mr Blunt's desiderata list.

Notebook, mainly travel and expenses.

Calendar of Lepidoptera, 1798–1827.

Insects contained in the collection of C. W. Dale and not in collection of J. C. Dale, 1871.

Itinerary of travel to the north.

Catalogue of all orders of British insects in the cabinets of J. C. Dale, with map and key to coloured labels.

Cut copy of Curtis's *Guide* indicating species taken by J. C. Dale and also in whose collection species may be found (based on 1829 ed.).

Capt. C. Blomer's entomological journal from 1820 to 1835 and catalogue of Blomer's cabinet.

Notes on contents of letters received and acted upon by J. C. Dale.

C. W. Dale entomological calendar, 1859–1905.

Lepidopterist's register of species in collection of C. W. Dale.

C. W. Dale catalogue of British insects added to the Dalean Collection, 1864–1905.

Catalogue of Hymenoptera in collection of C. W. Dale; Coleoptera, 1886; Aptera, Hemiptera, Homoptera, Neuroptera, Orthoptera and Hymenoptera Tenthredinidae, 1886; Diptera, 1886; Plants, 1886.

Catalogue of shells in the collection of M. L. Dale and Admiral Hancock; also catalogue of birds' eggs in C. W. Dale collection.

C. W. Dale catalogue of Lepidoptera Rhopalocera and Heterocera.

Correspondence: the letters number over 5000 and the total number of his correspondents is about 287 (excluding societies etc.), and includes almost all the entomologists of his day. J. C. Dale carefully numbered every letter he received and they are mostly in their original envelopes. Particular mention must be made of J. Curtis, whose letters number around 350, H. Doubleday 209, R. S. Edelston 118, A. H. Haliday 120, A. H. Haworth 36, F. W. Hope 11, W. Kirby 15, F. O. Morris 112, F. Walker 73, R. Weaver 184, and J. O. Westwood 10. C. W. Dale had letters from 37 correspondents.

The Hope file contains letters from J. C. Dale, 1823, 1824.

Westwood's file includes lists of insects sent to Ashmolean Museum and Oxford Museum; 'Coleoptera ex Insula Ste Helenae desiderata Mus. Hopeiana Oxoniae'; letter to P. B. Duncan, Ashmolean Museum, 1838, and two sets of photographs to Westwood, 1862. There are also four letters from C. W. Dale to Westwood.

Typescript of notes by S. C. S. Brown on the Dale diaries, letters, library, and collection, copies of wills of J., J. C., and C. W. Dale, and other miscellaneous papers.

DARBISHIRE, C. J.   Letter to Hope, 1830, with list of Coleoptera.

DARWIN, C.   Letters to Hope, 1833, 1837; note by Hope on *Hydroporus Darwinii*; copy of letter from Hope to Darwin, 1834.

Four letters to Westwood, 1860–1; letter to R. H. Wedgwood, 1885; copy of letter from Darwin to (?E. A. Ormerod, 1868); letter to J. Jenner Weir, 1871; thirty-five letters and eight postcards to Meldola, 1871–82 (no. 33 missing); one to A. Russel Wallace, 9 July 18—.

DARWIN, F.   Three letters and one postcard to Meldola, 1882; letters to Poulton, 1889–96.

DARWIN, L.   Letters to Poulton, 1929 and 1935.

DAY, C. D.   Seven albums of photographs of entomologists with biographical details; catalogue of British Chalcids in his collection; some notebooks; correspondence, 1933–69, including letters from B. S. Doubleday and his son John Day.

DEJEAN, COMTE P. F. M. A.   Letters to Hope, 1822, 1823; list of insects; notice of sale of insect collection with names of purchasers inserted by Hope; sale catalogue of his library, 1840.
   Fragment of manuscript (Westwood MSS).

DENNY, H.   Letters to Hope, 1839, 1842; letters from Westwood, 1842, 1868; list by Westwood of Denny's Anoplura, 1871, manuscript notes, pencil sketches, fifteen proof plates, and copper plates for his 'Monographia Anoplurorum Exoticorum' (?unpublished) purchased 1871. Letter to Westwood 18 July 1845 pasted in *Monographia Anoplurorum Britanniae* states 'you certainly deserve to be considered Quarter Master General to the Exotic Anoplura for your being always upon the look out to provide for my wants'.

DEYROLLE, A., H., and E.   List of 'Insectes de Colombie vendus a Monsieur Hope par Deyrolle' and 'Catalogue de Coléoptères de la Nouvelle Hollande'.
   Letters to Westwood, 1855–87, and various catalogues.

DIXEY, F. A.   Four notebooks of observations, 1871–1934; two notebooks with details of captures, 1905–32; two notebooks on Mortehoe Lepidoptera, 1879–1900; twenty-five notebooks, 1893–1933; lantern slides prepared for meeting of British Association, Sydney, 1926; two albums and two small boxes of pencil sketches of scent scales; diagrams illustrating mimicry, protective coloration, and seasonal variation, 1888–1919; correspondence.

DONISTHORPE, H. ST J. K.   Notebooks: 'A list of species of British Coleoptera not taken by me up to June 30th 1923; noted in red when taken', vols I–III; 'The Coleoptera of Windsor Forest Park and District'; Windsor Collecting; Collecting Journals 1–3, 1929–43.
   Manuscript notes 'Coleoptera of Windsor Forest' I–V; annotated copies of *Catalogue of British Coleoptera* by Beare and Donisthorpe, 1904; *A catalogue of the recorded Coleoptera of the British Isles* by Beare, 1930, and interleaved copy (with manuscript notes) of *A preliminary list of the Coleoptera of Windsor Forest*, 1939.
   Group photograph taken in front of the Watch Oak in Windsor Great Park October 1929 in a large carved oak frame. Note on the back states 'The beautifully carved Windsor Oak frame was originally presented to Prince Christian with an illuminated address, "1831–1911. To HRH Prince Christian of Schleswig Holstein from the Crown Officials and Employees in Windsor Great Park on his eightieth birthday. Jan. 22. 1911".'

DONOVAN, LIEUT.-COL. C.   Letters, notes, and original paintings; map of South India, 1924, giving localities for his butterfly labels.

DONOVAN, E.   '28th Sept. 1857. Received from Mr. Hope. A large collection of drawings of fossil Crustacea and Trilobites by Donovan with MS notes & impressions in sealing wax of the fossils' (entry in Westwood's journal). The Archives also contain other original drawings and paintings of fossils including a large number of fossil plants, and according to a manuscript note by Westwood 'Donovan's Rough MSS'. These include 'Miscellaneous observations & notes for Elements of Entomology' which was never published and the following manuscripts which are of especial interest:

'Alphabetical list of the larvae received from Mr. Fichtel & Mr. Francillon with the numbers added under which they are to be found in Ent. Syst. of Fabricius.'

'List of the larvae of Lepidopterous insects that I purchased of Mr. Fichtel & Mr. Francillon . . .'

'Neuroptera. Marsham's Insects.'

'Coleopterous Insects named in Marsham's Cabinet—now in my possession.'

Various notes on Marsham's Insects, Haworth's Lepidoptera, etc.

'List of Marsham's Insects in my possession. Hemiptera.'

'List of the Dipterous Insects in my possession that are named by Mr. M. Marsham's Insects.'

'Extract from Leach's work on entomology called Samouelle's.' (*The entomologist's useful compendium*, 1824.)

'Objections made by Leach on the supposed errors of Marsham and others . . .'

Original paintings and plates, chiefly exotic Lepidoptera.

Drawings from Jones' Icones with some manuscript notes by Westwood.

Bound collection of drawings and paintings of butterflies copied from Jones' Icones (many of which were published by Donovan in his *Naturalist's repository*, 1823-34) which also contains many paintings by Westwood (in Hope Library).

Dissected copy of *The natural history of British insects*, 1792-1813.

Dissected copies of first editions of *An Epitome of the natural history of insects of China*, 1798, and of *An epitome of the natural history of the insects of India*, 1800-3, used by Westwood in the preparation of his second editions of these works.

Plates from *The natural history of British fishes*, 1802-8.

Some correspondence, including two letters from William White 24 and 26 October 1893 to Miss Swann, who appeared to have offered to sell Donovan's drawings and manuscripts to the Ruskin Museum, Sheffield.

Dos Passos, C. F.   Letters to Carpenter, 1947-9, and list of species of *Papilio* donated to Collections.

Doubleday, B. S.   Manuscript notes, notebooks, and correspondence; *see also* Day, C. D.

Doubleday, E.   Manuscript lists of part of Libellulidae and Tortricidae in British Museum Collection and notes on various butterflies (Westwood MSS).

Douie, Lady Mary   Original painting of Indian insects, 1898-1900 (presented 1925).

Drury, Dru   Four notebooks of exotic insects in his collection [*c*.1784] from which E. Donovan prepared sale catalogue. They were subsequently given to Donovan as he declined payment for the work. Advertisement of Sale, London, May 1788; sale catalogue May 1805 (annotated with names of purchasers and prices).

Ecklon, C. F.   Letter to Hope, 1834, and account for South African insects purchased by Hope.

Edelston, R. S.   List of 'Lepidoptera in R. S. Edelston's Cabinet 20. 1. 45' and letters to J. C. Dale 1840-55 (Dale MSS).

Edwards, F. W.   First draft of 'The generic names of British Diptera Nematocera' (Verrall-Collin MSS).

Eltringham, H.   Original paintings and drawings; portrait of Professor Poulton taken through the eye of a glow-worm (in Poulton file); scrapbook; lecture notes in typescript; some correspondence.

Erichson, W. F.   Letters to Hope, 1834-9, including lists of insects.

ESCHSCHOLTZ, J. F.   Letters to Hope, 1827–31; for account of collection of Coleoptera offered for sale by his widow *see* letter to Hope from H. M. Asmuss, 1836.

ESENBECK, C. G. NEES VON   Thirteen 'Drawings of Chalcididae & Proctotrupidae from Esenbeck's typical Collection belonging to the Museum of Bonn' (Westwood manuscript note), and a few other notes by Westwood. Some of the drawings are coloured (Westwood MSS).

EVANS, F. C.   *see* VEVERS, H. G.

EVANS, H. SILVESTER   Manuscript notes and correspondence with Poulton, 1927–37, regarding material sent from Africa, also map showing collecting centres. A few notes have been cut up and pasted into one of Poulton's notebooks.

EVANS, R.   Letter to Poulton, 1899; diary of Skeat Expedition to the Siamese Malay States, 1899 (2 vols) and manuscript notes on butterfly collection.

EVANS, W. F.   'A bibliographical account of the works and writings on Entomology: Published in this country since 1574', unpublished manuscript (Westwood MSS).
   Manuscript list of Libellula at sale of his collection made by J. C. Dale, lots 30–33 [22 November 1870] (Dale MSS).

FABRICIUS, J. C.   'Manuscript of Fabricius. Given to me at Kiel 27th August 1835 by one of his two sons Dr. Fabricius . . . J. O. Westwood.' (one sheet of botanical notes) (Westwood MSS).

FEISTHAMEL, BARON J. F. P. DE   Letter to Hope, 1829, and desiderata list; two letters to Westwood, 1850.

FFENNELL, D. W. H.   Diaries, 1940–77; index to entomological notes from 1940 onwards.

FITCH, A.   List of insects sent to Westwood, 1852, and exchange list of New York insects.

FLOWER, —   *see* CLARK, J. W.

FORBES, C.   Letter to H. W. Acland, 1866, concerning collection of Vancouver Crustacea being sent to Oxford.

FOSTER, LIEUT. C. A.   Notebook of captures, 1910–13.

FRAMPTON, REVD R. E. E.   Letter to Poulton, 1926, and list of butterflies presented to Collections.

FRASER, LIEUT.-COL. F. C.   Manuscript notes, locality lists, drawings, including some of Indian Papilionine larvae, and correspondence, 1937–59; coloured plates of British Orthoptera originally intended for book in Wayside and Woodland series. Letters, mainly from K. J. Morton; *see also* MORTON, K. J.

FRENCH, G. H.   List of Lepidoptera from Illinois (in Ormerod file).

FRIVALDSZKY, E.   Letters to Hope, 1837–45, including lists of insects and 'Catalogue of Insects given me [Hope] by Julia Pardoe, and by her received from Fridwalsky the Entomologist'.

GEBLER, F.   Letters to Hope, 1828–31, with lists of Coleoptera.

GELDART, W. M.   Notebook, 1904–12.

GENÉ, C. G. [GENÉ, J.]   Letters to Hope, 1835, 1837, 1844, and lists of insects from Sardinia.

GÉNY, PH.   Original paintings of Crustacea, frogs, snakes, and lizards, also one sheet of fossil Crustacea (Hope–Westwood MSS).

GERMAR, E. F.   Letters to Hope, 1829–31, including lists of insects.

GISTL, J.   Letters to Hope, 1832–4, 1860, and Westwood, 1860. Two manuscript catalogues of his Coleoptera offered for sale at £40 and £100. When Westwood was Secretary of the Entomological Society he received a letter sent to Hope in 1839 from A. F. Schaesler, the Bavarian Consul in London, regarding Dr Gistl, who was at that time in prison in Munich 'under prosecution for having fraudulently contracted debts, without having at the time any reasonable prospect of being able to satisfy his creditors', amounting to £500 sterling. Gistl had apparently told the authorities that he had an agreement with the President of the Society under which he would be paid £3 per printed sheet, and they wanted an affidavit to this effect. Westwood replied repudiating any such agreement.

GOFFE, E. R.   Card index of references to his collection of Diptera especially Syrphidae; correspondence concerning bequest, 1952.

GORHAM, REVD H. S.   Letter to Westwood, 1882.
Notebooks and miscellaneous manuscript notes, including catalogue of species and genera of Coleoptera described by him since 1873; album of photographs of well-known entomologists; two small flower paintings and a few letters.

GORY, A.   Letter to Hope, 1849, offering his brother's (H. L. Gory) collection for £300; also letter to J. G. Children, 1837.

GORY, H. L.   Letters to Hope, 1828–37; lists of insects; list of entomological library; manuscript notes on 'Observations sur les cetoines que Monsieur Hope m'a communiquées' and 'Caractères des genres de notre Monographie Melitophiles de Latreille'.

GRAVENHORST, J. L. C.   Letters to Hope, 1828, 1830–8, including list of insects.

GREGORY, J. W.   Original painting for frontispiece of *The Great Rift Valley*, 1896.

GUÉRIN-MÉNEVILLE, F. E.   Letters to Hope, 1837–57; list of insects sent to Hope, 1841; offer of a collection of originals from 'Iconographie du Règne Animal' for £600 and collection of exotic Coleoptera for £80, 1841; letters to Westwood, 1842–55, 1861 (the 1861 letter is bound in copy of *Histoire Naturelle des Crustacés*, 1832).

GYLLENHAL, L.   'Catalogue des Insectes envoyés à Monseigneur le Baron De Jean, Juin 1820' (Westwood MSS).
See also SCHÖNHERR, C. J.

HAAN, W. DE   Letters to Hope, 1830–42, 1850 (copy), including lists of insects sent from Rijksmuseum van Natuurlijke Historie, Leiden.

HAINES, F. H.   Lists of Araneae, Phalangidae, Chernetidae, etc., collected by Haines, and list of abbreviations used for the data in his collection; some letters from A. R. Jackson, 1930–4, and one from Haines to O. Pickard-Cambridge, 1901.

HALIDAY, A. H.   Note concerning specimens to Hope, 1846.
Letters to Westwood, 1836 (specimens to Entomological Society) and 1863–9.
Manuscript notes entitled: 'Fam. Thripsidae'; 'Arrangement of Muscidae Acalypterae'; 'Haliday's remarks on the very minute Clypeaster from the *Deserta Grande*'; 'Monogr. of Thripsidae'; '5 drawings by Haliday'.
Two drawings and descriptions of *Pulex canis* in Westwood's notes labelled 'Aphaniptera'.
List of material in Haliday collection prepared by W. F. de Vismes Kane.
Notes and drawings by Westwood, and correspondence on Haliday collection from E. P. Wright, W. F. de Vismes Kane, and J. Steele, 1882–3.
For original drawings and manuscript notes prepared by Haliday for *Insecta Britannica. Diptera, see* WALKER, F. (Westwood MSS).

Visiting card of 'Lieu^t Col^l Haliday, 36th Regiment' taken from Dale letter from Haliday no. 53 [brother].
   Letters to J. C. Dale, 1841–69 (Dale MSS). *See also* McLachlan, R.

Hancock, Admiral   Catalogue of shells (mainly marine) (Dale MSS).

Hanitsch, K. R.   Manuscript notes including list of species described by him and location of the Types; notebook on Orthoptera including Blattidae of South-East Asia, and checklists of Blattidae; notebook 'Types of Blattidae described by K. R. Hanitsch'; original drawings.

Harwood, B. S. and P.   Diaries, 1888–1916, 1918–57; notebook on malaria and sleeping sickness; correspondence concerning bequest.

Harzer, A.   Manuscript list of duplicates sent to Hope (no date).

Haworth, A. H.   Manuscript list of Coleoptera in Mus. Haworth made by J. O. Westwood, 1823; letter to Westwood, January 1828; and two sheets of the *Lepidoptera Britannica* (Westwood MSS).
   Letters to J. C. Dale, 1820–8 (Dale MSS).

Hearsey, Maj.-Gen. Sir J. B.   Entomological and botanical diary, 1838–46 and two original paintings (Westwood MSS).

Heath, R. H.   Six notebooks on Lepidoptera with records of captures.

Hewitson, W. C.   Outline sketches of butterflies apparently relating to his work *Illustrations of diurnal Lepidoptera*, 1862–78 (Westwood MSS).
   Some correspondence and miscellaneous notes by Westwood.

Holdsworth, E.   Letter to Westwood, 1866, containing description of larvae of 'Arctias Luna' and original painting of caterpillar.

Holland, F. W.   Manuscript notes on 'Locust-eating Crickets', 1867 (Westwood MSS).

Hollick, A. T.   Original drawings and paintings, including a collection of illustrations, *c.*1870, for a Ray Society Monograph to supplement Blackwall's *Spiders of Great Britain and Ireland*, planned by O. Pickard-Cambridge but never published; some correspondence (O. Pickard-Cambridge MSS).

Holme, W. F.   Manuscript 'Catalogue of Coleoptera taken by myself in Gloucestershire'; letters to Hope, 1837–9.

Hope, Revd F. W.   Diplomata Hopeiana, 1832–49, and silver medal presented by the Duke of Tuscany.
   Notebook, copper plates, and 49 coloured copies of each of four plates (del. J. O. W.) for 'Coleopterist's Manual. Part the Fourth' (not published).
   Catalogue of Portaits of Naturalists, dating from 1565.
   'Mr. Hope's Collections' (Anatomical Museum Christ Church, 1852).
   Notebook: includes list of cabinets, visitors, 1840–1, and list of Coleoptera.
   Notebook: portraits, painters and engravers, naturalists, travellers, and miscellaneous notes.
   Notebook: 'Memoranda 1839' containing list of books.
   Notebook: list of correspondents, dealers, and general notes on insects including notes for 'Entomologia Sacra' etc.
   Manuscript notes 'Coleoptera of New Holland'.
   'Catalogue of Coleoptera begun in 1828' and six other notebooks on Coleoptera.
   Four paintings attributed to Hope.
   'Entomology of Oxford 1819'.

Index to Herbst's Coleoptera, also to Dalman, Klug's Century, Palisot, and Dejean's *Species général des Coléoptères*.

'Mr. Hope's tables of Insects attacking the Human Body'—Tables 1–6 mounted on five sheets.

Unpublished manuscript entitled 'A notice of some locusts which have ravaged India in the years 1843 and 1844 [observed] by Lieutenant H. B. Edwardes of Her Majesty's First European Regiment of Light Infantry'.

Notebook in which Hope has written on inside cover 'Dr. Bacon's Collection of [Indian] Hemiptera bought by J. O. Westwood'; it contains a numbered locality list for the Hemiptera, also Lepidoptera, Diptera, and Hymenoptera and could be in Dr Bacon's handwriting corresponding to numbers on some specimens in the Hemiptera collection.

Entomological notebooks (forty-three paperback and two hardback) containing notes for his work 'Entomologia Sacra', also 'Entomological Miscellany', 'African Entomology Vol. 1', 'Fauna Australasiae' (short list of insects), 'Insects taken at Cheltenham in 1844' (few pages), 'Remarks on the Modern Classification of Insects 1838'; list of 'Fabrician Curculionidae'; biographies; 'British entomological authors'; lists of British and foreign entomologists with name of painter, engraver, etc.; list of portraits of prelates, clergy, gentry, etc.

Manuscript notes: general zoology including fishes, birds, Crustacea etc.; entomology, including miscellaneous lists of insects; extracts from works on insects in amber.

Lists of Italian entomologists; lists of royalty, gentry, clergy, etc.

Catalogue de la collection de Coléoptères de M. le Baron Dejean, 1821 (2nd edn); interleaved copy, with manuscript notes by Hope, 'All Insects in this Catalogue marked √ thus are in the Collection of the Revd F. W. Hope'.

Catalogue des Coléoptères de la collection de M. le Comte Dejean, 1833–4 (3rd edn); parts 1–4 interleaved with manuscript notes by Hope. Also a copy of the second issue of the 3rd Edition profusely annotated by Hope (copies of both editions in Hope Library).

Academical *Éloges*, bibliographical and biographical notices, etc. of distinguished men of all countries formed by the Jussieu family: 'The loss of this collection has been much regretted by the Savan[t]s of Paris who had hoped that the Administration of the Jardin des Plantes would have secured them for that Establishment' (Westwood's manuscript note); also a similar collection from the library of Brongniart.

Collection of early pamphlets on chemistry, physiscs, meteorology, engineering, education, etc.; *The year-book of facts in science and art*, 1841–52, 1855, 1857, 1858, 1861–2; Clerk's *Naval tactics*, 1804.

Correspondence in connection with benefaction, and lists of items despatched to Oxford at various times.

'Catalogue of a Collection of Tracts . . . presented to the University of Oxford by the Rev.ᵈ Fredᵏ Wᵐ Hope . . . 1851. Scr. John Macray.'

Correspondents, apart from those already mentioned under the individual concerned, include L. A. Barthélemy, C. Bassi, G. C. Berendt, C. S. Bird, F. de Brême, M. Chaudoir, H. Cuming, T. Curties, S. H. De Saram, L. W. Dillwyn, C. F. Drége, L. Dufour, T. Eyton, F. Falderman, G. Fischer von Waldheim, J. Fleming, E. L. J. H. Boyer de Fonscolombe, C. D. E. Fortnum, J. Gould, G. R. Gray, L. Guilding, T. Hardwicke, M. Herold, W. F. Holme, W. J. Hooker, T. Horsfield, J. Kidd, — Kindale, W. Kirby, I. H. Lance, F. L. N. de C. de Laporte (Comte de Castelnau), J. E. LeConte, J. Macquart, C. F. Ph. de Martius, A. Melly, E. Ménétriés, G. Newport, R. Owen, F. J. S. Parry, T. J. Pettigrew, F. J. Pictet de la Rive, C. Robert, C. Rondani, J. F. Royle,

W. Sells, A. J. J. Solier, W. Spence, W. Stephenson, J. H. Stewart, S. L. Strachan, H. E. Straus-Dürckheim, H. E. Strickland, W. H. Sykes, G. Thorey, P. Walker, C. R. W. Wiedemann, and V. P. de Zoubkoff.

See letter from W. Bainbridge, 27 April 1841, for list of 'Named Species of Lucanidae in Mus. Rev. F. W. Hope of London'.

Hope Curators   Minute Book of the Hope Curators' Meetings from 1857 to 1944, and Hope Curators' accounts (also contains a copy of Deed of Gift and relevant correspondence, made by Westwood) and 1945 onwards. These minute books also contain records and cuttings from *University Gazette*, etc.

Horn, —   List of crustaceans collected in the Red Sea, sent March 1870.

Howitt, G.   Letter including manuscript list and observations on the insects sent to Hope, 1842.

Hübner, J.   Slip cases containing figures cut from *Sammlung Europäischer Schmetterlinge*, mounted presumably by Westwood, and letters concerning purchase in 1852; 580 plates from his *Sammlung exotischer Schmetterlinge* purchased by Westwood in 1862, many having handwritten headings etc. [?by Hübner] (Westwood MSS).

Hunter, Maj.   Illustrations of transformations of Indian Lepidoptera painted *c.*1851–3 (unpublished) and two letters to Westwood, 1880 (Westwood MSS).

Image, S.   Five notebooks and miscellaneous manuscript notes.

Jackson, T. H. E.   Letters to Carpenter, 1944–52 (containing collecting information) and copies of his replies.

Jacobson, E.   Letter to Poulton, 1928, and list of insects sent from Sumatra.

Jermyn, Col. T.   Manuscript catalogue of collection of Indian Lepidoptera (presented in 1938 by Royal Entomological Society of London who possess his manuscripts and books).

Jones, W.   Seven volumes bound in six of paintings of Lepidoptera known as 'Jones' Icones' with short Latin descriptions, 1783–5 (Plate 15): Papiliones Equites [Vol. I]; Papiliones Heliconii [Vol. II]; Papiliones Danii [Vol. II]; Papiliones Nymphales [Vol. III]; Papiliones [Vol. IV]; Papiliones [Plebeii] [Vol. V]; Papiliones [Vol. VI]. Some of the species figured were described by Fabricius, and the paintings of these species are Iconotypes, since the original specimens are lost.

Lens, palette, and colours used for the 'Icones'; painting instructions; eighteen original paintings of Lepidoptera, Lepidopterous larvae, tulips, and *Auricula*.

Notebooks:

British moths and butterflies, 1780–5, and extracts from Drury's Entomological Diary, 1764–6;
One notebook on butterflies and two on moths;
British Lepidoptera, 1770–90, and a copy of this notebook;
Botanical notes;
Historical dates and notes;
Commonplace book (receipts, flowers in bloom, etc.);
Rough copy of 'Minute Book of the Society of London Entomologists, 1780–1782'.
Four notebooks of verse, chiefly sacred;
'Directions for illuminating M.S. Transcribed from a M.S. of 1710 by Mrs. Elstob';
Two manuscript copies of 'An Introduction to the Hebrew Language with A Short View of the Hebrew Words in English by John Mackay 2 September 1767';

'Brachygraphy or Short-Writing' by Tho. Gurney; and another notebook on short-writing.

Manuscript notes:

'Duties of the Converted towards their Neighbour';
Translations from Greek and Latin (sacred);
Notes on colony of moths;
Descriptions of caterpillars, their food plants, emergences, etc.

'The Speech of Logan, a Shawanese Chief to Lord Dunmore, Governor of Virginia in 1774.'
Pedigree of George III from Wodin and Friga.
Letters from Dr Smith (Sir James Smith owner of Linnaeus' collections) from Paris, September 1786, Rome, February 1787, and Genoa July 1787; J. Tetlow, February 1798; T. Marsham, October 1789 (fragment).

Copies of:

A dissertation on the sexes of plants, J. E. Smith, 1786;
A catalogue of the plants growing wild in the environs of London [W. Curtis], 1774;
A concise Hebrew grammar, by W. Romaine, 1803.

Bound volume containing:

Instructions for collecting and preserving insects, by W. Curtis, 1771;
Fundamenta Entomologiae, by W. Curtis, 1772;
Institutions of Entomology, by T. P. Yeats, 1773.

Framed silhouette of William Jones by T. Rider.
Westwood's manuscript notes on Jones' Icones.
Manuscript of article by F. D. Drewitt 'Illustrations of Exotic Butterflies described by Fabricius from the Drawings of Jones', 1871, also some proof plates, probably not published; notes by Westwood, and letters from Westwood to Drewitt.
Annotated interleaved copy of A. H. Haworth's *Prodromus Lepidopterorum Britannicorum*, 1802.
These manuscripts were presented by F. D. Drewitt, 1925–33.
See also DONOVAN, E.

JORDAN, R. C. R. and W. R. H.   List of Lepidoptera in Museum Jordan, Teignmouth, March 1841, by number only (Dale MSS).

JOURDAIN, REVD F. C. R.   Two notebooks containing locality lists.

KENT-LEMON, A. L.   Letters to Poulton, 1918–20; two maps of collecting areas, and Poulton's notes for data labels.

KIRBY, REVD W.   Letters to Hope and Westwood, 1822–38.
Letters to J. C. Dale, 1818–27 (Dale MSS).
Sheet of 'Sketches of dissections of parts of mouths of Dynastidae and Melolonthidae' (Westwood MSS).

KLUG, J. C. F.   Letters to Hope, 1828–42, including lists of insects.

KOLLAR, V.   Letter to Hope, 1845, and manuscript lists of Coleoptera and Hymenoptera.

KUNZE, G.   Letters to Hope, 1832, 1836, and lists of insects.

KUNZE, R. E.   Letters to Poulton, 1910, 1913, and from his nephew Richard Kunze, 1929, with list of butterflies.

Lack, D. and Venables, L. S. V.   Venables' letter to Carpenter (no date) and data for insects collected on Galápagos expedition.

La Ferté-Sénectère, F. de   'Anthicites communiqués par Mr. Hope.'

Lamborn, W. A. S.   Lists of species of insects and plants collected in East Africa, Kuala Lumpur, etc.; correspondence between Poulton and Lamborn, with manuscript notes and photographs of butterflies, 1910–32; official correspondence 1910–23.

Lampel, G. P.   Diaries and notebooks, 1955–60, and correspondence.

Lankester, C. H.   Letters to Poulton, 1921–6, 1933–4, concerning donation of insects; letters to Carpenter, 1932, 1937.

Lankester, E. R.   Letters to Poulton, 1883–99, 1906, also many undated. Cartoons executed at meetings: Board of Faculty 1891; Committee for unassigned Zoological Collections at Oxford, 1893; Delegates of Museum accepting report of Committee for unassigned Zoological Collections, 1893.

Latreille, P. A.   Letters to Audouin, 1830 (offprint 50669), Baron de Férussac, 1830, Westwood, 1830, and note introducing him to L. A. de Brébisson; two sheets of manuscript notes. Manuscript notes made by Westwood on Latreilles's genera of Proctotrupidae and Chalcididae, also pencil sketch of Latreille (Westwood MSS).

Leach, W. E.   List made by J. C. Dale of specimens in cabinet in British Museum (Natural History), March 1844, mainly Odonata (Dale MSS).

Lefebvre, A.   Letters to Hope, 1831–7; to Westwood, 1851–2; manuscript on wing venation by Lefebvre and Westwood.

Lemon, A. L. Kent-   *see* Kent-Lemon, A. L.

Linnaeus, C.   List of Libellula in Linnaean Cabinet, 1846 (Dale MSS); details of statue of Linnaeus presented to University Museum by Hope; copper plate of Linnaeus; one sheet from *Systema Naturae* (1st edn), 1735: IV Pisces, V Insecta, VI Vermes (now hanging in Library).

Longstaff, G. B.   Letters to Poulton, 1899–1912, 1919; notebooks and lists of captures, 1903–20; records of butterflies captured by Capt. R. S. Wilson in the Nuba Mountains, Sudan; original paintings and drawings for *Butterfly hunting in many lands*, 1912; drawings for *Bionomic notes on butterflies*, 1909.

Lucas, W. J.   Letters to Poulton, 1898, 1904, 1906. Original paintings for: plates 1–27 of *British dragonflies (Odonata)*, 1900; plates 4–35 of *The aquatic (naiad) stage of the British dragonflies (Paraneuroptera)*, 1930; plates 3, 8, 11, 13, 17, 21, 22, and 23 of *A monograph of the British Orthoptera*, 1920; other paintings and drawings, some published, and a few photographs.

Lupton, Miss P. M. and Wykes, Miss U. M.   Card index and manuscript notes, Iceland Expedition, 1935.

McComish, Mrs I.   Some correspondence concerning her Norfolk Island collection and copy of the British Museum (Natural History) preliminary report.

McLachlan, R.   Correspondence with Westwood, 1862–86.
   27 boxes of letters; correspondents (approximately 800) include many well-known naturalists, such as T. D. A. Cockerell, A. E. Eaton (numerous letters and lists of captures), H. and P. H. Gosse, E. E. Green, H. A. Hagen, A. H. Haliday, J. J. F. X. King, F. Klapálek, J. Lubbock, L. R. Meyer-Dürr, K. J. Morton, F. Müller, R. Oberthür, C. R. Osten-Sacken, F. P. C. A. Preudhomme de Borre, C. V. Riley, F. Ris, E. de Selys-Longchamps, H. T. Stainton, O. Staudinger, and T. Wiltshire.

Manuscripts and drawings, including those for McLachlan's article on Neuroptera, 1875 (in Fedtschenko, A. P., 1874–80, *Reise in Turkestan*, II Zool. Theil (7), 60 pp.). The McLachlan letters and manuscripts were presented by Colonel F. C. Fraser, *c*.1959.

See MORTON, K. J. *and also* WALLACE, A. RUSSEL.

MACLEAY, W. S.   Letters to Hope, 1839–43, to Westwood, 1833; list of Annulosa sent to Hope in 1842.

MANNERHEIM, COMTE C. G. VON   Letter to Hope, 1830, with list of Coleoptera.

MARSHALL, REVD T. A.   Letters to Westwood, 1868, 1884–5; two original paintings.
'Some of account of T. A. M[arshall]'s earlier days'; letters from E. Saunders to Poulton, 1905, concerning collection.

MASON, C.   Data supplied by Lamborn for Mason's Nyasaland collection.

MATTHEWS, REVD A.   Letter to Westwood, 1886, with list of Trichopterygidae.

MEADE, R. H.   List of British spiders purchased at Entomological Society Sale, July 1863; letters to Westwood, 1888.

MELDOLA, R.   Letters from and to Poulton, 1884–1901, 1911, 1913, and from A. Weismann, 1878–84; entomological notebook and 'notes to accompany my collection of moths'.
See DARWIN, C. *and also* WALLACE, A. RUSSEL.

MIERS, J.   Four catalogues of insects in his collection, each catalogue containing sheets of manuscript notes, the last with a letter from Carnier, 1860; letters from his son J. W. Miers, 1880, and Westwood's lists of the collection; notes by grandson, Professor H. J. Miers, to Poulton, 1896.

MILLAR, H. M. and BLAKE, J.   Correspondence, 1954–5, concerning collection and list of Lepidoptera presented.

MOORE, M. S.   Letter to Poulton, 1937; field notes and list of insects.

MORICE, REVD F. D.   Letter to Westwood, 1881; letters to Poulton, 1911, 1918, and 1921, from Longstaff to Morice, 1910 and 1914; letter and notes from Perkins to Poulton concerning collection, 1928.
Notebooks and diaries, various notes, lists, photographs, and a few drawings, also includes personal photographs of his family, E. Frey-Gessner, Arnold Friese, and J. H. Comstock.
Letters from eighty-six correspondents including J. D. Alfken, E. André, P. Blüthgen, R. du Buysson, T. D. A. Cockerell, E. Enslin, H. Friese, E. Frey-Gessner, A. Handlirsch, F. W. Konow, A. Mocsary, R. C. L. Perkins, J. Pérez, S. A. Rohwer, O. Schmiedeknecht, A. V. Schulthess-Schindler, W. Trautmann, and A. Russel Wallace.

MORTON, K. J.   Letters, including many from F. C. Fraser and R. McLachlan. *See* FRASER, F. C. *and also* MCLACHLAN, R.

MOYSEY, MAJ. F.   Letter to Carpenter, 1935, and locality list of butterflies from southern Sudan.

MULSANT, E.   Letters to Hope, 1850, 1852, and Westwood, 1850, 1863.
List of Hope specimens used for his *Species des Coléoptères trimères sécuripalpes*, 1850.

NATIONAL ANTARCTIC EXPEDITION   Official documents and correspondence, 1901–4, concerning Joint Antarctic Committee (Royal Society and Royal Geographical Society) and appointment of an Executive Committee on which E. B. Poulton was a representative of the Royal Society.

Neale, J. P.   Original drawings (coloured) of insects, 1807–[1809], in notebook; also three separate plates from *Trans. Ent. Soc. Lond.*, 1807–9 (Hope–Westwood MSS).

Neave, S. A.   Letter to Poulton, 1910?; letters from A. V. Schulthess-Schindler, 1922, concerning determination of Neave's Rhodesian Hymenoptera.

Nees von Esenbeck, C. G.   *see* Esenbeck, C. G. Nees von

Newman, E.   Letters to Hope, 1833 (includes subscription list concerning Stephens v. Rennie) and to Westwood, 1829, 1857, 1863, 1868.

Norwich Entomological Society   Minutes of third meeting held at Mr Curtis's house, 4 December 1810.

Nuttall, G. H. F.   List of Ixodoidea presented to Poulton.

Oates, C. G.   Letters to Westwood, 1877–88; Westwood's manuscript notes for entomology section, in Oates, F., 1881, *Matabele Land and the Victoria Falls*, edited by C. G. Oates (Westwood MSS).

Obert, J.   Letter to Hope, 1842, with list of Siberian Coleoptera.

Ocskay von Ocsko, Baron F.   Letters to Hope, 1835–7, includes list of Orthoptera.

O'Mahony, E.   Notebooks: 'Systematic notes on Mallophaga and Siphonaptera etc.' and 'Records of beetles in Co. Dublin (North East)'.
   Manuscript: 'A catalogue of the Siphonaptera Collection and Siphonaptera of Ireland' (two volumes, in Library).
   Some correspondence.

Oxford Microscopical Society   Minutes of meetings, 1864–70.

Oxford University   Miscellaneous documents and letters to Westwood concerning Ashmolean, Pitt-Rivers, and University Museums, 1857–84; also letters and documents concerning University affairs; pamphlets and letters to Poulton, 1900–1.

Oxford University Entomological Society   Minute books, 1858–72, 1922–78; correspondence concerning *Accentuated list of British Lepidoptera*, 1859; two letters from H. T. Stainton presenting volumes of *Natural history of Tineina*, 1858–9; letter from Secrétariat de l'Université Royale de Norvège (June 1863) concerning Mr P. A. Munch; 'Cantate ved det Kongelige Norske Frederiks Universitets Mindefest Hans Majestaet Kong Carl den 19de November 1872'.

Oxford University Expeditions   Manuscript notes Greenland, 1928; typescript lists Lapland, 1930; manuscript list Hudson Strait, 1931, and Greenland, 1936; some correspondence.
   List of Coleoptera from Arctic Norway attributed to Oxford University Expedition.

Palisot de Beauvois, A. M. F. J.   Original painting for plate XX, *Insectes recueillis en Afrique et en Amérique*, Paris, 1805–21, '. . . given to me October 1850 by Dr. Boisduval. Jno. O. Westwood' (Westwood MSS).

Parfitt, E.   'Larva Book' containing manuscript notes and original paintings; manuscript of *Fauna of Devon Hymenoptera. Section Aculeata*, 1880, and of *Fauna of Devon Phytophagous Hymenoptera—Sawflies* 1888 (L. H. Woollatt bequest).

Parreyss, L.   Letter to Hope, 1834, and lists of Coleoptera.

Pascoe, F. P.   Scrapbook (mainly cut-up plates of beetles, but including a few original paintings); original paintings, also a few pen and ink, and pencil sketches; three notebooks; some correspondence. *See also* Wallace, A. Russel.

Passerini, C.   Letters to Hope, 1834–53, and lists of desiderata; list of insects sent to Westwood; letter to Boisduval, 1830.

Passos, C. F. Dos   *see* Dos Passos, C. F.

Pearce, Miss E. K.   Correspondence concerning her collection including letters to Poulton; annotated copies of *Typical flies*, series I–III, 1915–28; notebook and list of Diptera; biographical details.

Pecchioli, V.   Letters to Hope, 1834–44, and desiderata lists.

Peile, Col. H. D.   Three lists of butterflies sent to Oxford, 1948–9; original painting of *Daphnis nerii* caterpillar.

Perkins, R. C. L.   Letters to Poulton and Carpenter, 1890–1946; notes on sawfly cabinet and F. Smith cabinet; list of Hymenoptera sent to Oxford, 1946; letter to Hobby, 1949, with list of some of the Aculeates sent to Oxford.
   Manuscript notes, a few drawings, and list of sawflies in Perkins collection (Morice MSS).

Perris, E.   Fragments of manuscripts (Westwood MSS).

Pickard-Cambridge, Sir A. W.   Manuscript notes mainly on spiders and Lepidoptera; letters from Wm. Falconer and J. E. Hull, 1918, and R. Dalmas, 1920; letters to Poulton, 1920, 1930, 1937; correspondence concerning bequest, 1952–9.

Pickard-Cambridge, F. O.   Miscellaneous notes, mainly on spiders; his interleaved copy of *Spiders of Dorset* by O. Pickard-Cambridge, 1879–81, with manuscript notes and drawings; some correspondence.

Pickard-Cambridge, Revd O.   Catalogue of spiders in his collection; original drawings and paintings, notes, and letters regarding spiders, European and exotic; various notebooks; proofs of plates for Blackwall's *A history of the spiders of Great Britain and Ireland*, 1861–4; report on Blackwall's Type specimens received through T. West; exotic spiders described by Blackwall; list of Types of European Theridiidae from Kulczyński; Salticidae determined by Peckham and Pickard-Cambridge and manuscript list of Salticidae; note on manuscript of Albin's spiders; letters from more than 70 arachnologists, including Blackwall, Koch, Kulczyński, Meade, Peckham, Simon, and Thorell.

Poulton, E. B.   Correspondence with Westwood, 1880–92.
   Manuscript notes, photographs, and original paintings for published work, of particular importance being the notes on *Papilio dardanus*, including breeding experiments and relevant correspondence, and those on predaceous insects and their prey.
   Notebooks referring to material given to Hope Department; correspondence during tenure of Hope chair including letters from Poulton to Carpenter, 1910–29. *See* Trimen, R. *and also* Wallace, A. Russel.
   Photographs, including portrait of Poulton taken through the eye of a glow-worm by Eltringham, 1918; personal mementoes.
   Typescript copy of letter signed by Lankester, Poulton, and Romanes entitled 'An Institut Transformiste' containing extract from lecture by M. Giard.
   Terms of E. B. Poulton 'Evolution Fund'; *see also* Baldwin, J. Mark.

Putzeys, J. A. A. H.   Hope's manuscript list, 'Clivina Dyschirius &c e Musei Dom. Hope 1845 Insecta transmissa Dom. Putzeys Bruxelles'.
   Letter to Westwood, 1864.
   For notes on Types of *Clivina*, *see* Andrewes, H. E.

Raddon, W.   Letters to Hope, 1836, 1841; letter to Curtis, 1838, with list of North American Coleoptera extracted from turpentine.

Manuscripts entitled 'Synopsis of the North American species of *Hymenoptera* collected by Mr. Raddon Esq., from Turpentine' and 'Descriptions of some North American Coleoptera obtained from raw Turpentine by W. Raddon Esq.' by Westwood, apparently unpublished. Subsequently one or two descriptions were published in his *Thesaurus* and the *Zoological Journal* (Westwood MSS).

RÉAUMUR, R. A. F. DE   Sheet of manuscript notes '1729 8er Mars histoire naturelle' given to Audouin by Cuvier, 1829, and by Audouin to Westwood, 1837 (Westwood MSS).

REICH, G. C.   Letters to Hope, 1833–45.

REICHE, L.   Letter to Hope, 1842, and list of 'Insectes offerts à Monsieur Hope par Reiche Juin 1844'.

REID, COL.   *see* CLARK, J., 1854.

RICHARDS, O. W.   List of Hymenoptera with prey (presented 1937).

RISSO, A.   Bound volume of paintings entitled 'Crustacés de Nice'; paintings for his posthumous work on Cephalopods (37 plates) (Hope–Westwood MSS).

ROBINSON, A.   Four sheets of diagrams of fossil insects, *c*.1896.

ROE, J. S. and W.   Letters to Hope, 1828–39 (W. Roe) and 1833–44 (J. S. Roe); 'Notes on the Insects of Mr. Roe's Collection taken in 1828' by Hope and colour-coded locality list from W. Roe, 1829.

ROGERS, K. ST AUBYN   Letters to Poulton, 1908–17, and Carpenter, 1936, containing collecting information.

ROLLESTON, G.   Photograph of Dr Rolleston's class, 1858, taken by Charles L. Dodgson (Lewis Carroll) in Dr Lee's Museum, Christ Church.
Letters to Poulton, 1876–80.

ROTHNEY, G. A. J.   Notebook 'Register book of Captures (with rough notes) Aculeata Hymenoptera', 1863–94; three large scrapbooks of letters, notes on his collection, lists of insects, localities, photographs, etc.; correspondents include: E. André, W. Ashmead, P. Cameron, A. Forel, G. Mayr, C. G. Nurse, E. B. Poulton, E. Saunders, S. S. Saunders, F. Smith, and J. O. Westwood.

ROWLAND-BROWN, H.   Notebooks, 1887–1921; catalogue of Lepidoptera, 2 volumes; typescript lists of Lepidoptera and manuscript catalogue of Zygaenidae; letter from W. S. Gilbert, 1911, and one from Rowland-Brown to Dixey, 1912; letters from his sister to Dixey and Poulton, 1922–3, and one from C. Oberthür to sister.

SAHLBERG, C. R.   Letters to Hope, 1830–5; 'Index Insectorum ad Cel. *Dom.* Hope missorum' and 'Catalogus Insectorum "fennia et Lapponia" qua' Helsingforsia venduntur . . . '.

SALE CATALOGUES   Series dating from 1786 to 1980; includes a small bound collection which belonged to E. Donovan, and annotated copies relating to Francillon, Dru Drury, Haworth (Library), Children, Doubleday, Entomological Society, Edwin Brown, J. A. Turner, etc.

SAMOUELLE, G.   Fragment of 'Samouelle's MSS attempt to edit Drury 2nd Edition' (MS note by Westwood) (Westwood MSS).

SARG, F. G.   Original paintings; letters to Salvin, 1881–2; notes on Araneidea from Guatemala, 1881–3 (O. Pickard-Cambridge MSS).

SAUNDERS, E.   Correspondence with Westwood, 1867–89, and Poulton, 1896–1908; diary, 1878–1909, giving details of his captures.

SAUNDERS, SIR S. S.   Letters to Westwood, 1852–84 (Westwood MSS).

Three notebooks entitled 'Entomological Notes'.

Strepsiptera manuscript, 1871, 'Strepsipteridarum, pro Ordine Strepsipterorum Kirbii olim, mihi tamen potius Coleopterorum Familiae, Rhipiphoridis Melöidisque propinque, Monographia. Auctore S. S. Saunders.'

Saunders, W. W.   Letter to Hope, 1842; letters to Westwood, 1851–79, and other correspondence and notes concerning purchase of his collection; letter from C. Horne, Trinidad, to W. W. Saunders, 1866; letters from G. F. Hampson and S. Stevens to Poulton concerning collection, 1895–6; two notebooks with numbered records of his captures.

Savage, T. S.   Letters to Hope, 1840–8, and to Westwood, 1844–8, containing much information concerning specimens collected.

Savory, T. H.   Notebook concerning his collection.

Say, T.   Letter to Hope, 1828, expressing his intention to send insects on his return from Mexico; letter to C. R. W. Wiedemann, 1830.

Schaum, H. M.   List by Westwood of insects received from North America, 1848; letters to Westwood, 1853, and to Wollaston, 1854.

Schaus, W.   Letters to Poulton, 1894–c.1909, 1925; manuscript list prepared by Schaus for *On Walker's American types of Lepidoptera in the Oxford University Museum*, 1896.

Schiödte, J. M. C.   Letters to Westwood, 1854, and list of insects sent.

Schönherr, C. J.   Letter to 'Monsieur le Comte', July 1827; letters to Hope, 1829–46, including lists of insects, particularly Curculionidae. Manuscript notes, some in L. Gyllenhal's handwriting (Westwood MSS).

Serville, J. G. Audinet   Letters to Hope, 1834–41; notice of death, 1858.

Shelford, R. W. C.   Letters to Poulton, 1907, 1911, 1912; also to Shelford from various correspondents including Bolivar, Burr, Gestro, Horvath, Rehn, etc.; a few pencil sketches and manuscript notebook containing some sketches and many loose notes.

Sidgwick, A. and N. V.   'List of moths and notes upon them', 1853–6; entomological diaries, 1857, March 1860, 1880–1913; miscellaneous manuscript notes and some correspondence.

Silbermann, G. H. R.   Letters to Hope, 1824–35, and list of Coleoptera.

Smith, F.   Letter to Westwood, 1854; list and correspondence with Janson, 1879, concerning purchase of Hymenoptera. *See also* Perkins, R. C. L.

Smith, H. Grose   Letters to Westwood, 1890–1, including lists of specimens purchased.

Smith, H. H.   Two volumes of 'Notes on spiders and other alcoholic collections', by H. H. Smith, 1887–8, Mexico; manuscript notes on 'Alcoholic Specimens' made by H. H. Smith, Orizaba, Mexico, December 1889 (O. Pickard-Cambridge MSS).

Smith, W. G. Pogson   Notebook 'Entomological Calendar January 1896–1912'.

Smithsonian Institution, US National Museum   Letter from Assistant Secretary to Westwood concerning collection of birds' eggs from Hudson Bay Co., sent *via* Smithsonian Institution, 1870; also manuscript data mostly by Westwood.

Westwood's list of marine invertebrates received from US National Museum, 1884.

Snellen Van Vollenhoven, S. C.   Letters to Westwood, 1860–71, and lists of insects.

Society of London Entomologists   For rough draft of Minute Book, 1780–2, *see* Jones, W.

Sommer, M. C.   Letter to Hope, 1835, concerning Coleoptera from Dr K. C. A. Zimmermann.

Spilsbury, Revd F. M.   Letters to Westwood, 1878–80, from Revd B. W. Spilsbury concerning bequest of his brother's collection and extract from will.

Spinola, Marchese M.   Letters to Hope, 1835–44, and list of 'Insectes offert au Revd W. Hope par Maximilian Spinola Mai 1844'.

Stainton, H. T.   Letters to Westwood, 1853–87, and manuscript 'The Neglected Orders'.
'Novitates Staintoniana, or catalogue of British Lepidoptera added to the collection of H. T. Stainton Esq. in the year 1842. Consisting of 103 species' comprising 81 pages of manuscript and original paintings. Presented 1912 by Mrs P. B. Mason.
For letters from Stainton to McLachlan *see* McLachlan, R.

Stephens, J. F.   Letters to Westwood, 1826, and to Hope, 1835–7. Part of Stephens's manuscript for *A systematic catalogue of British insects*, 1829 (Westwood MSS).
Documents in connection with Stephens *v.* Rennie court case and copy of Rennie's book used as evidence during the hearing of the case and subsequent arbitration.

Stevens, J. C. and S.   Details of purchases and letters, 1858–89.

Stewart, [C.]   'Stewart's Manuscript 1802' on Coleoptera (incomplete) (Westwood MSS).

Stockley, Col. C. H.   Correspondence with Carpenter, 1947–52, lists of butterflies sent to Oxford, and a few photographs.

Strong, W.   Letters to Hope, 1837 from his daughter Esther Strong concerning sale of collection.
Manuscript notes, and paintings by Westwood (Westwood MSS).

Sturm, J.   Letter to Westwood, 1824, and catalogue of insects for sale, October 1829.
Letters to Hope, 1830–1, with several manuscript lists of insects.

Sutton, G. P.   Two paintings illustrating, but not published in, *A remarkable brood of Aglais urticae*, 1943.

Swainson, W.   Letters to Hope, 1828–33; sketch of coral found near Naples by Mr Pitt, 1803.
Letters to Westwood, 1830–3, concerning his contribution to 'Encyclopaedia of Zoology'. Manuscript notes and impressions of woodcuts by Westwood (Westwood MSS).

Swynnerton, C. F. M.   Manuscripts (some in typescript) of unpublished articles on form and colouring, attacks by wild birds on butterflies, and feeding experiments with lemur and cat; some correspondence, including letters to Poulton, 1910–11, 1926.

Sykes, Col. W. H.   Letter to Hope, 1830; manuscript list of 'Coleoptera from India in the Museum of Col. Sykes. 1833' by Hope.

Talbot, G.   Manuscript notes on African Lycaenidae by Talbot, who revised the Oxford Collection; copies of photographs of new forms described in *Bulletin of the Hill Museum*.

Templeton, R.   Unpublished manuscript 'Figures and descriptions of Irish Arachnida and Acari', nos. 1–6 [1834–49?] (O. Pickard-Cambridge MSS).

Original drawings for *Description of a minute Crustaceous animal from the Island of Mauritius*, 1840; proof plates 16 and 17 for *Memoir on the genus Cermatia*, 1843; manuscript note 'Mr. Templeton's Elenchus' and some original drawings; a few letters to Westwood, 1854, concerning culture of silk, also some letters with Westwood's 'Genera Insectorum' (Westwood MSS).

THAYER, A. H.   Two typescript copies of letters to A. Russel Wallace, *c.*1905; 14 copyright photographs, many with manuscript notes on reverse.

THOMAS, R. H.   Manuscript list of specimens with localities, collected in Ecuador, 1937–8; letters to Carpenter, 1937–9.

THWAITES, G. H. K.   Letters to Westwood, 1871–2, 1877, 1881, and E. Stainforth Green, 1877, containing much information about fig insects and other material.

TRIMEN, R.   Notebook, 1856; a few miscellaneous notes and photographs for a paper 'On the larvae of *Hamanumida daedalus*, Fab., *Hoplitis phyllocampa*, n.sp.', 1909; letters from Trimen to Poulton, 1886–1915, and Poulton to Trimen, 1904–10.
Award of Darwin Medal and letters of congratulation, 1910–11.
Correspondence, 1902–12, including some letters from E. Bourke, H. Eltringham, Miss M. E. Fountaine, G. A. K. Marshall, A. D. Millar, S. A. Neave, and L. Peringuey. (Most of the letters contain a copy of Trimen's reply.) Other personal notes.

TURNER, H. J.   Correspondence concerning his collection, 1917, 1935, 1949.

TURNER, J. A.   Letter to Hope, 1829, sending specimens.

TYLECOTE, E. F. S.   List of European specimens made by Carpenter, 1933; some correspondence concerning collection.

VARNEY, W.   Letter to Westwood, 1852, with manuscript notes and observations on the natural history of various insects (published in *Trans. Ent. Soc. Lond.* **1853**) (Westwood MSS).

VENABLES, L. S. V.   *see* LACK, D.

VERKRÜZEN, T. A.   Letters to Westwood, index, and list of species in collection (purchased 1888).

VERRALL, G. H. AND COLLIN, J. E.

VERRALL: Extensive collection of manuscript notes on Diptera; interleaved copy of *A list of British Diptera*, 1888, and 2nd edition, 1901, by Verrall; also his personal copies of volumes 5 and 8 of *British flies*, 1901 and 1908; 'Census of British Diptera in my collection, September 10th 1896'.
Racing diary with lists of horses participating in various events.

COLLIN: Original paintings and drawings for published work including those for Verrall's *British Flies* volumes 5 and 8, and his own volume 6.
Manuscript notes, notebooks, and index to the recorded genera and species of British Diptera.
Copy of first draft of 'The generic names of British Diptera Nematocera' by F. W. Edwards sent to Collin, July 1940 (presumably not published).
Letters, 1964–73, concerning gift of Verrall–Collin Collection.

Large number of letters from eminent Dipterists to both Verrall and Collin, including some from the following: J. M. Aldrich, H. W. Andrews, E. E. Austen, E. B. Basden, T. Becker, J. M. F. Bigot, E. N. Bloomfield, E. Brunetti, J. W. Carr, C. W. Dale, F. W. Edwards, E. C. M. d'Assis-Fonseca, A. H. Hamm, C. G. Lamb, H. Loew, W. Lundbeck, G. A. K. Marshall, C. R. v. Osten-Sacken, O. Parent, the Revd A. Thornley, C. J. Wainwright, F. Walker, F. A. Walker, J. H. Wood, and J. W. Yerbury.

VEVERS, H. G. and EVANS, F. C.   Field notes concerning collection from island of Myggenaes, Faeroes, July 1937.

VILLA, A. AND G. B.   'Insectes pour M.ʳ F. W. Hope de Londres Mai 1844.'

VILLIERS, F. DE   Letters to Hope, 1835–7; manuscripts of five works by Villiers.

VINALL, MISS G.   Letter to Poulton, 1930; some data concerning her collection.

WAIN, FATHER F. L.   Notes on his collection of Indian bees and typescript on nests of bees.

WALKER, F.   Letters to Hope, 1848, and Westwood, 1833, 1855.
Manuscript index to Deltoites, Pyralites, Geometrites, Noctuites, and Bombycites in *List of the specimens of Lepidopterous insects in the Collection of the British Museum*, 1854–66; original notes and drawings prepared by A. H. Haliday and Westwood for *Insecta Britannica. Diptera*, 1851–6 and unbound copy containing some manuscript notes by Westwood; manuscript description of Chalcidoidea from Japan prepared for F. Smith's article in *Trans. Ent. Soc. Lond.* **1874** (Westwood MSS).
List of Diptera 'In Mr. Walker's collection I have seen 22.1.71' by G. H. Verrall to which is appended the following note: 'Mr. Verrall advised me to destroy the insects as their cond^n was miserable. E. B. P.' and 'G. H. Verrall's list of F. Walker's Diptera, given to Hope Dep^{mt} Mch. 1906 when I sent the box of original specimens to him. It was evident that box had been dropped & insects & labels mixed up indiscriminately. All specimens were destroyed exc. Chilosia olivacea ♂ q.v. in Brit. Coll^n E. B. Poulton Mch 10. 1906.'
Typescript list of 'Walker's Aphids in Hope Dept. Collections Nov. 1952' prepared by J. P. Doncaster.
Letters to J. C. Dale, 1833–71; manuscript list of Chalcidoidea sent by F. Walker to J. C. Dale in 1847 (some of these are certainly syntypes of his species) (Dale MSS).

WALKER, J. J.   Letters to Poulton, 1896–1901.
Twenty-four notebooks covering years 1870–1911, containing information concerning captures in Great Britain and during his journeys overseas, including the Mediterranean area, 1875–8, outward and homeward voyage of HMS *Kingfisher* on Pacific Station, 1880–4, Gibraltar, 1886–9, Australia, Tasmania, China, Colombo, and Aden, 1890–2, Hong Kong, 1892–3, voyage to Sydney on P & O *Oceana*, 1899, New Zealand, 1901–2, Sydney to Hobart, 1903, expedition to Blue Mountains, 1903–4, and voyage home.
Photograph taken on board ship.

WALLACE, A. RUSSEL   Letters to Westwood, 1865–71.
Correspondence with E. B. Poulton, 1886–1913 (includes letter from C. Lloyd Morgan, 1891, and Charles Darwin, 9 July —, and one from Wallace to B. R. Miller, 18 January 1913); manuscript notes by Poulton extracted from Wallace letters, 1896–1913 (paste-up); letter from Ternate, 20 December 1860, to F. P. Pascoe; letters to R. McLachlan, 1871–90, R. Meldola, 1879–1910, and Countess of Warwick, 16 August 1900. *See also* THAYER, A. H.
Dates of Wallace's expeditions.
Manuscript by R. Meldola on A. Russel Wallace and his theories.
Copy of *The scientific aspect of the supernatural* by Wallace, 1866, with manuscript account on flyleaf by F. Sims (née Frances Wallace) of events leading up to 'spirit writing' appearing at top of flyleaf—this presentation copy from the author to his sister is only one of the many books forming the 'A. Russel Wallace Spiritualist Library' presented by T. H. Riches in 1915.

WALLACE, E. A.   Postcards to Westwood, 1884, 1886, and list of butterflies from Colombia purchased by Westwood.

WALLER, W. E.   Notes on Lepidoptera from Iraq and the Pyrenees, 1918–20.

WATERHOUSE, G. A.   Letter to Poulton, 1911, with information on Australian Pieridae sent to Oxford.

WATERS, E. G. R.   Entomological diaries, 1906–29.

WATKINS and DONCASTER   Letters to Westwood, 1881, 1885; invoices, 1887, and lists of specimens purchased by Westwood.

WEISMANN, A.   Letters to Poulton, 1914; manuscript notes relating to *Studien zur Descendenz-Theorie*, 1875–6; letters to R. Meldola, 1878–84.

WESMAEL, C.   Letter to Westwood from T. A. Marshall, 1885, enquiring about location of Wesmael Types of Braconidae.

WESTERMANN, B. W.   Letters to Hope, 1827–44, containing lists of insects sent to Hope and also 'Some remarks on the Carabi, described by Fabricius . . . now in the Collection of the Royal Museum at Copenhagen' with letter dated 1839.

WESTWOOD, J. O.   *Diplomata Westwoodiana*, 1827–89.

Sixty-five boxes of manuscript notes, drawings and paintings, etc., so extensive that only a small portion could be enumerated below; these include copious manuscript notes, drawings and paintings for his own published work, and notes and abstracts from published works by other authors. The arrangement has, as far as possible, been kept as Westwood left it.

Diaries and notebooks

Diary covering years 1820–37; notes made during visits abroad: Paris, 1830 (with 'Notes on Chalc^{ae}. & Proctotrup^{ae}.' and pencil sketch of Latreille), 1837, 1842, 1844, 1850, 1852, 1860, 1865; Kiel, Berlin, etc., 1835; Copenhagen, Stockholm, etc., 1869. 'Entomologia; sive Excerptorum, auctoribus è variis Scientiae illae, Collectio. Tom: 1ers Johanne O. Westwood. Excerpta. 1823'; also three similar manuscript notebooks dated 1825–33.

'Generum Insectorum Index Methodicus, continens (si completus)'.

Insects (including Crustacea and Arachnida): Surveys and Arrangements, four volumes.

General entomology

'Genera Insectorum': a systematic work in manuscript in nineteen original slip-cases, profusely illustrated with drawings and paintings, including drafts of published works and letters from eminent entomologists such as Haliday, McLachlan, W. W. Saunders, Spinola, Templeton, and G. R. Waterhouse.

Manuscript of title page and preface for proposed work entitled 'The Entomological Miscellany', *c*.1827–8, which was apparently never published. This manuscript was found in the same bundle as the manuscripts of two articles entitled 'Descriptions of some North American Coleoptera obtained from raw Turpentine by W. Raddon, Esq.' and 'Synopsis of the North American species of Hymenoptera collected by W. Raddon, Esq., from Turpentine' by Westwood which were perhaps intended for the 'Entomological Miscellany'; subsequently one or two descriptions were published in his *Thesaurus* and the *Zoological Journal*.

'J. O. W. Entomol. MSS & Drawings—for Swainson's Cyclop: Nat. Hist^y &c. unpublished'; this was part of a proposed work entitled 'Encyclopaedia of Zoology' edited by Swainson.

Thesaurus Entomologicus Oxoniensis, 1874; manuscript notes, paintings, drawings, and sketches for the plates and correspondence relating to the work.

J. Curtis's economic and taxonomic notes, drawings and paintings, with many additions by Westwood.

Manuscript for entomological section of C. G. Oates's *Matabeleland and the Victoria Falls*, 1881.

'Insectorum Familiarum (Historiae Naturalis) Index'.

Notes and impressions of woodcuts for *Introduction to the modern classification of insects*, 1838–40.

Notes for articles in *Magazine of Natural History*, 1829–32.

Notes from Curtis and Melly Collections, 1841.

'Observations on the economy & transformations and habits of various insects made in 1837'.

Lists of 'Insects &c. in Spirit'.

'MS. Nov. Sp. Descripsit. Hope' and 'N. Sp. Ins^{ct} Descripsit. Westwood'.

'Objections to the System of Macleay'.

'Aptera': manuscript notes and drawings, also letters from O. Pickard-Cambridge, H. Ridley, T. Whitmarsh, A. Müller, and T. A. Chapman.

'Drawings of mites found on Almond . . .' from Mr F. Crawford, Adelaide, to Miss Ormerod, October 1881.

'The Tick from Canterbury Cathedral' by G. Gulliver, *c*. February 1872.

'MS New Introd^{n} to Entomology'.

'Monstrosities'.

Fossils

Manuscript notes and drawings on fossil insects; also letters from P. B. and W. R. Brodie, 1853–4, and plates from published work.

Crustacea and Arachnida

Manuscript notes, drawings, and proofs for *A history of the British sessile-eyed Crustacea* by Bate and Westwood, 1863–8, including letters from Bate, Milne-Edwards, T. Q. Couch, P. H. Gosse, and H. Powell; agreements of 1854 and 1857 between Westwood and Van Voorst, the publishers.

Drawings and lists of Crustacea, notes on Swedish and Italian Crustacea, letters from J. Thompson, W. V. Cocks, W. Sanford, A. Jewitt, A. R. Hogan, J. Byerley, and R. Patterson.

Miscellaneous notes, drawings, and plates, and abstracts from other works; notes and drawings from publications by J. D. Dana.

Original drawings, notes, and proof plates for Bell's *A history of the British stalk-eyed Crustacea*, 1853.

Manuscript notes and drawings of Arachnida and 'A note on the impregnation of Arachnida', 1828.

Plates by Westwood from published work.

Hemiptera, Strepsiptera, and smaller orders

Manuscript notes, drawings, proofs, and plates including: genera of Heteroptera and Homoptera, *Coccus*, *Cimex*, *Fulgora*, and *Derbe*; Stylopidae brought by S. S. Saunders from Albania, 'Notes on Hylecthrus pupa' and 'Mr. Templeton's Elenchus', with original drawings by Templeton, and letters from J. W. Douglas, G. H. Ford, S. Green, R. McLachlan, S. S. Saunders, E. Shepherd, V. Signoret, and W. Spence; E. Doubleday's manuscript 'Libellulidae in British Museum Collection';

Thysanura; Ephemeroptera; Psocoptera; Thysanoptera; Neuroptera; Trichoptera; Aphaniptera.

Dermaptera and Orthoptera

Manuscript notes, original drawings, and some plates, mainly on Orthoptera.

'Ten drawings of Achetidae exotic species I. O. Westwood Feby. 1874'.

Lists of grasshoppers, locusts, and crickets described by L. H. Fischer.

Drawings of specimens seen in various collections—Hope, Guérin, Banks, Lefebvre, Marchal, Jardin des Plantes, and Mus. Paris.

Notes and drawings for his *Revisio Insectorum Familiae Mantidarum*, 1889, and *Catalogue of Orthopterous insects in the collection of the British Museum. Part I, Phasmidae*, 1859.

Lepidoptera

Manuscript notes, drawings, etc., some for published work including Saturnidae, Bombycidae, Hesperiidae, *Castnia*, Uraniidae, and transformations.

Bound collection of drawings from Jones' Icones by Donovan and Westwood (in Hope Library).

'List of species of Papilio described by Fabricius from Jones' Icones not repres[d] in any of Donovan's subsequent works'.

List of 'Original drawings of Butterflies made by Donovan now in Hopeian Collection at Oxford or copies from Jones, etc.'.

E. Doubleday's list of part of Tortricidae in British Museum Collection.

'Principles of classificat[n] of Lepidopt.'.

Sketches and notes on *Himantopterus* and *Thymara*.

Notes, paintings, and sketches on 'Kirby's Exotic Moths', including letters from W. F. Kirby.

Abstracts and notes from published work other than Westwood's.

Notes and lists of diurnal Lepidoptera, including 'Catalogue of Part of the Diurnal Lepidoptera of the Jardin des Plantes 1850—Nymphalidae & Hesperidae', and 'Boisduval's Nymphalidae, Hipparch[ae] & Erycinidae Oct. 1850', copied from Boisduval's cabinet.

Plates from published work 'J. O. W. del.'

Coleoptera

Manuscript notes and drawings, some for published work, including work on Paussidae (with letters from J. G. Children, R. Brown, and Dr Horsfield and copies of letters from Westwood); *Cremastocheilus* and Lucanidae (letters from F. J. S. Parry and H. Schaum); a few drawings of Wollaston's Madeira Coleoptera; Coleoptera from Hong Kong and China with list by Major J. G. Champion, and 'Description of a new genus of Carabideous insect from the Upper Amazon, Brazil'.

Papers on *Staphylinus* and other genera of families of Coleoptera.

'Instrumenta Labialia of various Coleoptera'.

'Copies of drawings made for Andrew Murray's Monograph', 1859–60.

Interleaved copy of *Catalogue de la collection de Coléoptères de M. le Baron Dejean*, 1821, with manuscript notes by Westwood (in Hope Library).

Collection of plates from published work 'J. O. W. del.'

Diptera

Manuscript notes and drawings mostly for published work including notes on genera of Diptera, Muscidae, Cecidomyidae, Mydasidae, Oestridae, Streblidae,

Nemestrinidae, Tipulidae, Acroceridae, etc.; letters from Osten-Sacken and one from Ingpen.

Plates from published work 'J. O. W. del.'

See also WALKER, F. for original drawings and manuscript notes prepared for *Insecta Britannica. Diptera.*

Hymenoptera

Manuscript notes and drawings including some for published work, particularly on fig insects and Chalcididae, Proctotrupidae, Cynipidae, Bethylidae, Dryinidae, Eurytomidae, Ichneumonidae, etc., and containing letters from W. T. Dyer, W. Fowler, T. A. Marshall, S. S. Saunders, F. Walker, C. O. Waterhouse, and draft of letters to J. Curtis; also manuscript entitled 'The Bee'.

'Notes for Hist.y of Trichiosoma & parasites'.

'Habits & Transformations of Hymenoptera'.

Drawings of *Nematus gallicola* and notes on gallforming sawflies etc., and their parasites.

Notes on Nees von Esenbeck collection and thirteen drawings of Chalcididae and Proctotrupidae from his typical collection belonging to the Museum of Bonn.

Manuscript notes and extracts from works by Dalman, Fabricius, Fonscolombe, Geoffroy, Huber, Jurine, Latreille, Spinola, and Walker.

Plates from published work 'J. O. W. del.'

Economic entomology

Manuscript notes and drawings, mainly for articles on pests including queries to *Gardener's Chronicle*, and letters from J. S. Henslow, A. H. Haliday, W. J. Hooker, and Miss Ormerod, also some to Loudon's *Gardener's Magazine.*

Bees and beehives with illustrations of early beehives.

List of injurious insects.

Bethnal Green Branch of the South Kensington Museum, Committee of Advice and Reference on the Economic Entomology Collection. Minutes, correspondence, 1864–86, and 'Scheme for a series of Illustrations of Economic Entomology by Professor Westwood M.A., F.L.S.'

Culture of silk, including letters from W. Reid to R. Templeton, and from Templeton and E. Solly to Westwood.

Cocoons and sample of silk and part of silk series purchased 1888; piece of ox's hide pierced by more than 100 warbles, presented by Miss E. A. Ormerod, 1888.

Collection of paintings and sketches by Westwood

This collection of published and unpublished work includes:

Insects and Crustacea, *c.*1820–9.

Orthoptera (including one of a type from Curtis's cabinet).

British insects (some 1826–7).

Coleoptera—some species described by Hope and Westwood but not figured, *c.*1830.

Orchids, etc., including letter from W. W. Saunders sending orchid for inclusion in *Arcana.*

Sketches of Longicorn beetles made in Paris.

Gum insects in 'Mus. W. Strong' (including notes).

Illustrations for works published by Hope, W. H. Sykes, and G. R. Waterhouse.

181 demonstration posters prepared for teaching purposes.

Collection of plates by Westwood

From published work by T. Bell, P. B. Brodie, F. W. Hope, A. Murray, W. W. Saunders, J. F. Stephens, J. O. Westwood, and T. V. Wollaston.

Lecture notes

'Tsetse lecture'.
'Significance of Analogy in Natl Histy'.
'Reflections', 'Insects in General, etc'.
'On a hitherto unnoticed modification in the respiratory apparatus of certain Crustacea', British Association, Oxford, 1832.
'On the metamorphoses of insects', Royal Institution of Great Britain, 1861.
Lecture notes, 1862–91; Clarendon Press School, 1859, 1860.
Notes and correspondence for a 'Course of four lectures on Entomology', London Institution, 1867.
'On the data afforded by the Burchellian Collection', Ashmolean Society, 1866.
'Honey bees, wasps and ants', Ashmolean Society, 1880.

Ivories

Notices in *Oxford Times* and *Oxford Undergraduate Journal*, and notice of meeting of Oxford Architectural and Historical Society, 1872.
Sale catalogues of Italian and Flemish carvings formed by R. Goff, and objects of art and vertu collected by H. Farrer, sold June 1866.
Copies of published work including:
Catalogue of the Fejérváry Ivories in the Museum of Joseph Mayer, by F. Pulszky, 1856; contains letter from author, 17 April 1854.
Ein Kunst-Reliquie des Zehnten Jahrhunderts, by J. T. and P. S. Käntzeler, with translation and letter from W. Chaffers, 1856.
Ivory carvings, by Westwood, in *Stereoscopic Magazine*, 1859.
Diptychs of the Roman consuls, talk given by Westwood in Trinity Term, 1862, with annotations.
Early Christian art, illustrated by ivory carvings.
'. . . *objects pertaining to religious worship, from Russia . . .*', exhibited by Westwood.
Ancient and mediaeval ivories, by H. Clark, 1875.
Historische Austellung Kunstgewerblicher Erzeugnisse zu Frankfurt-a-Main, 1875.
Das Diptychon Quirinianum zu Brescia, by F. Wieseler, 1868.

Correspondents, apart from those already mentioned under the individuals concerned, include E. André, J. Bigot, W. Cutter, W. L. Distant, A. Dohrn, C. A. Dohrn, H. Donckier, H. Druce, F. Enock, A. Fauvel, W. Ferguson, J. E. Fischer von Röslerstamm, E. L. J. H. Boyer de Fonscolombe, J. C. Galton, H. H. Gigliolo, H. Goss, P. H. Gosse, J. E. Gray, S. Green, A. Grote, H. A. Hagen, T. W. Harris, S. Hickson, W. J. Hooker, F. W. Hope, F. J. Horniman, W. Houghton, W. H. Jackson, H. Jekel, J. T. Lacordaire, F. L. N. de C. de Laporte (Comte de Castelnau), J. L. LeConte, J. Lichtenstein, V. Lopez de Seoane, J. Wood-Mason, G. Mayr, F. Moore, J. Murie, A. Murray, C. and R. Oberthür, E. A. Ormerod, C. R. Osten-Sacken, F. J. S. Parry, J. A. Power, C. V. Riley, R. H. F. Rippon, C. Ritsema, H. de Saussure, S. H. Scudder, W. Spence, C. Swinhoe, W. Tylden, J. Waghorn, C. O. Waterhouse, J. P. M. Weale, R. C. Wroughton, and W. Yarrell.
Entomological equipment and personal mementoes.

WHITEHILL, COL. S.   Letters to Hope, 1831–7.

Whitmarsh, Revd T.   Letters to Westwood, 1871–87; small notebook and some loose notes; anatomical sketches and notes on *Synergus, Neuroterus, Rhodites, Alaptus,* and *Eupelmus.*

Wiggins, C. A.   Letters to Poulton, 1906–12; maps and manuscript notes; annotated copy of *On a large collection of Rhopalocera from the shores of the Victoria Nyanza* by Neave, 1904; catalogue of his mimetic series, 1909–13.

Williams, C. B.   Migration records, manuscript notes, and correspondence.

Williams, H. B.   Letters to Carpenter, 1945.

Wilson, C. A.   Lists of insects, sent 1849, 1853–4, and letter to Westwood, 1852.

Wilson, J.   Letters to Hope, 1827–35.

Wilson, Capt. R. S.   Letters to Poulton, 1917–23; manuscript notes by Poulton and Hamm concerning collection from the Sudan and Egypt.
Notebook prepared by Longstaff listing butterflies collected by Captain Wilson, also maps of White Nile and Sudan (Longstaff MSS).

Winckworth, H. C.   Notebook 'Winckworth collection of butterflies of South India and Ceylon. List of species. Index to localities'.
Transfers of scales of butterflies from the Andaman Islands.
A few letters concerning collection.

Winthem, W. von   Letters to Hope, 1829–35.

Wollaston, T. V.   Letters to Hope and Westwood, 1860–77, and papers relating to purchase of insects and shells; 'Reference to the insects in Madeiran Collection' [Coleoptera catalogue], purchased by Hope, 1860; manuscript lists of Madeiran and Canarian land shells, also Coleoptera from St Helena, received 1878.
Correspondence (Dale MSS).

Woollatt, L. H.   Manuscript notes including keys and determinations of insects in his collection by C. D. Day, R. B. Benson, E. A. Fonseca, J. E. Collin, G. M. Spooner, and R. C. L. Perkins; card index and miscellaneous photographs; correspondence, many with draft replies.
Photographs of 'saws' taken from Morice negatives by Woollatt (Morice MSS).

Yerbury, Col. J. W.   Letters to Poulton, 1896–1922; diaries, 1882–1926; catalogue of specimens; two notebooks on Diptera and some loose notes.

Zimmerman, K. C. A.   Letter to Hope, 1835, concerning American insects. *See also* Sommer, M. C.

Appendix B. The Collections and their donors

During the period when Westwood was active, both he and Mr and Mrs Hope acquired several large collections which comprised not only insects and arachnids, but also crustacea, reptiles, and various other zoological specimens, including even some birds' eggs. Hence it has been found advisable to mention the non-entomological items for the sake of producing a more complete account, although they were subsequently distributed to the appropriate sections of the Museum.

The especial aim has been to document all the older collections including small donations and purchases, details of which are not very well known (some information regarding these has been abstracted from Westwood's journal). The prices paid, when available, have been quoted for the sake of interest. Amongst more recent donations only the larger items are listed. Members of staff have at various times made considerable additions to the British Collections, especially to groups in which they were interested. Details of these, together with those of smaller gifts too numerous to mention, may be obtained by referring to the annual reports which have been published since 1883.

The list, which has been prepared from available records, is not exhaustive because of the absence of accession books between 1904 and 1932, and the complex way in which early departmental reports were presented, but it gives a fair representation of what the collections contain. The known date of accession is given as accurately as can be determined, or the date when it was first mentioned in a departmental report. This information is given in parentheses at the end of each item, and (n.d.) where the date is not known. The original forms of the localities have been retained since they are thus labelled in the collections; translating them into their modern equivalents would be impracticable.

In view of the increasing interest taken during the past few decades in the insect collections of the earlier classical period, it has been thought useful to draw particular attention to those contained in the Hope Collections. A few are remnants of collections otherwise largely or entirely destroyed, for example those of Nees von Esenbeck and Boyer de Fonscolombe. Specimens were carefully labelled, individually or in blocks, by Westwood at the time he acquired them, but were partially incorporated with the general collection, where they remained unnoticed until about 1952. Because of their importance, they were subsequently numbered, fully labelled, and transferred to special drawers in the Type Collection, so that they may be easily located.

It is fitting to note here Poulton's intent for the vast collections acquired from world-wide localities. He wanted the Collections to be a centre for ecological studies, and I quote from a letter dated 24 July 1927 he sent to Carpenter in Entebbe:

His [H. W. Simmonds] coll. has determined me to do what I have long thought of & I think you will approve—viz. Keep these Island Colls. together permanently as units. Then one can so easily test the changes that take place in the future. Also I think the Hodson Abyssinian Coll. Abyssinia is almost like an island in the N.E. of the continent & it is most interesting to see the forms which occur there, together.

In the main, accessions have been incorporated in the general collections, except for certain notable acquisitions some of which are now housed in the Historic Room, but many others still remain as discrete entities. It is to be hoped that the important historic aspect of the Hope Entomological Collections will not be lost with the present trend of amalgamation.

Key to some old labels in the Collections taken from slips pasted in Westwood's Journal:

Amaz } Bart }	The Amazon, Bart[lett]
Burnett R.	Burnett River
Calif. } Har }	California [Har or Hav]
Cantor	Collected by Dr. Cantor on Prince of Wales Island or adjacent islands (but some possibly from the continent of India or China)
Canys or Cany	Canary Islands
Cape Camb	Cape, Campbell
Captain Boys	India
Carth } Brewer }	Carthagena (Spain), Brewer
Colum. Chest.	Columbia, Chesterton. W. W. Saunders Coll.
Granad } Brewer }	Granada (Spain), Brewer
Haw	Haworth Coll. mainly white labels
Kas.	Khasyah [Kasia] Hills, India
M.I.	Melville Island, north of New Holland [Australia]
Madag.	Madagascar
Maurs } Desjs }	Mauritius, sent by J. F. Desjardins to J. O. Westwood
Mend Reed	Mendoza, Reed
New Cal.	New Caledonia
N. Scot	Nova Scotia
Or Exp	Oregon Expedition
Port Wel	Portugal, Welwitsch
Red R	Red River, North America
Singa	Singapore (pale blue label)
Tang } Brewer }	Tangiers, Brewer
Up. Amaz Bart.	Upper Amazon, Bart[lett]
Urug	Uruguay (blue label)
W	Coll. Westwood (blue lozenge-shaped label)

'Insects with old labels with a thick black line are from Lee's Collection of Hammer-smith often named by Fabricius himself.'

'Mexican insects with n^{os} refer to Mr. Coffin's ins[ects] from Coll^n. Westwood'

'Colours made use of in the Longicorn Collection on French Labels
   Peacock blue = Africa   Senegal, Cape, Guinea, Ile de France [Mauritius]
   Bright Yellow = Asia   Java, Arabia, Himalayas, East Indies
   Bright Green = America, N. & S.   Louisiana, Brazil
   Pink = Australia
   White = Europe   Constantinople'

A. Russel Wallace

Amb.	Amboina
Aru	Aru Is.
Bac.	Batchian
Bali	Bali Is.
Ban.	Banda Is.
Banca	Banca Is. (E. of Sumatra)
Bou.	Bouru
Cel.	Celebes
Cer.	Ceram
Coup.	Coupang (W. end of Timor)
Del.	Delli (E. end of Timor)
Dor.	Dorey (N. Guinea)
F. } Flo.	Flores Is. (W. of Timor)
Gag.	Gagie (small island W. of Waigiou)
Gil.	Gilolo
Gor.	Goram Is. (E. of Ceram)
J.	Jilolo or Gilolo Is.
Jav.	Java
Kai.	Kaioa Is. (N. of Batchian, Moluccas)
Ké	Ké Is. (E. of Banda)
Lom.	Lombok
M.	Mysol Is. (N. Guinea)
Mac. } Mak.	Macassar (S. Celebes)
Maki.	Makian Is. (W. of Gilolo)
Mal.	Malacca
Mal. P.	Malay Peninsula
Mat.	Matabella Is. (E. of Goram)
Men.	Menado (N. Celebes)
Mor.	Morty Is. (Gilolo)
N. or N.G.	New Guinea
S } Sal.	Salwatty (N.W. of N. Guinea)
Sar.	Sarawak, Borneo
Sing.	Singapore
Sul.	Sulla or Sula Is. (E. of Celebes)
Sum.	Sumatra
Ter.	Ternate

Tid.	Tidore Is. (S. of Ternate)
Tim.	Timor
Ton. } Tond. }	Tondano (mountain region of N. Celebes)
Wag.	Waigiou Is.

ADAMS, A.   Collection of Crustacea (dry) chiefly small species from the Eastern Ocean and Japanese Sea (voyage of HMS *Samarang*, 1843–6), and about 150 small bottles of species in spirits (from Mr Higgins, April 1872, £12. 10s.).

ADAMS, B. G.   British Lepidoptera (1912–14).

ADAMS, F. C.   Diptera from Lyndhurst district in the New Forest (1901, 1903, 1912).

ADDISON, MRS M.   Various insects from Sierra Leone (1911–12).

ADKIN, B. W.   122 Lepidoptera from Scilly, Orkney, and Shetland islands, etc. (1925).

ALLEN, G. DEXTER   Lepidoptera and Coleoptera from Switzerland and France, also specimen of *Limnas chrysippus* from Greece (1897).
   Oriental insects of many orders (presented by Mrs M. E. Dexter Allen, 1930).

ALLEN, G. O.   Indian butterflies, 881 set specimens, also many in papers (1939).

ALLEN, H. W.   315 Tiphiidae (Hymenoptera) mainly from the United States (1967).

ALLEN, W. R.   Lepidoptera from Florida and British Columbia (presented by H. P. Allen, 1902).

ANDERSON, A. W.   Butterflies from the Gold Coast (1927).

ANDREWES, H. E.   Coleoptera including syntypes of many new species (1915–22).
   Books and bound separata on Coleoptera (chiefly Carabidae) (1945–6).

ANDREWES, H. E. and F. W.   Collection of Coleoptera from India, Burma, New Guinea, Natal, Tennessee, including some syntypes of Jacoby, Horn, and Régimbart (1900).

ANDREWES, H. L.   Lepidoptera from Dar-es-Salaam and Lumbwa, Kenya Colony (1915–22).
   3648 Indian moths mainly from Nilgiri Hills (1932, 1935).

ANDREWS, C. W.   Miscellaneous insects from Christmas Island collected 1897–8 (presented by H. Druce, 1911).

ANGAS, G. F.   Insects from South Australia (purchased from S. Stevens, July 1863).

ANNANDALE, N.   Small collection of insects from Iceland and the Faröe Islands (1900).
   Insects from the Siamese States, collected by N. Annandale and H. C. Robinson, part of material described in *Fasciculi Malayensis* edited by Annandale and Robinson (1903).
   Insects from Malta and Faröe Islands (1904).
   14 Rhynchota named by Horvath, and eight Diptera named by Brunetti, Edwards, and Miss S. L. M. Summers, from Palestine (1913).

APLIN, O. V.   Butterflies from South America (1893).
   Butterflies collected in Uruguay, Tunisian Sahara, and arctic Norway (1896).

ARKELL, W. J.   British Diptera, principally Stratiomyidae and other Brachycera (1948).

ARNOLD, G.   Ants from neighbourhood of Bulawayo (many new species described by Forel), a few Australian and Californian ants, and a few South African Aculeates (1912).

South African ants (1915–22).
African Aculeate Hymenoptera by exchange (1941).
See also PERKINS, R. C. L.

ASH, REVD C.   Moths from Yorkshire localities (1923, 1925).

ASHBY, REVD E. B.   About 3000 butterflies from Britain and other parts of Europe (1936). (Manuscript notes by Ashby in his copy of *The butterflies of Switzerland* by G. Wheeler, 1903, in Library.)

ASHMOLEAN MUSEUM, OXFORD   Specimens from old Ashmolean Museum when collections transferred to New University Museum (1861). One specimen of *Gibbium psylloides* (Coleoptera, Ptinidae) found in 1800-year-old papyrus from Oxyrhynchus, Egypt, *c.* second century AD (presented by R. A. Coles, 1975).

ATKINSON, E. F. T.   European and Indian Hemiptera, chiefly Homoptera (1895, £5); contains many Lethierry specimens.

AUDOUIN, J. V.   Some syntypes of Chalcidoidea described by him in *Histoire des insectes nuisibles à la vigne*, 1842. No other syntypes hitherto located elsewhere.

AURIVILLIUS, C.   32 Lepidoptera, including eight butterflies collected in the Kilimanjaro district by the Sjöstedt Expedition, and a female of *Acraea medea* from Prince's Island, Gulf of Guinea (presented by S. A. Neave, 1911).

AUSTIN, E. P.   North American Coleoptera selected from his priced catalogue (August 1878, £4. 9*s.* 5*d.*)
North American insects (April 1879, 15*s.* 3*d.*).

BACKHOUSE, J.   Two specimens of *Papilio* from Ecuador (January 1869, 12*s.*).

BACON, DR   Collection of Indian Hemiptera *via* J. O. Westwood (1857); Lepidoptera, Diptera, Neuroptera, and Hymenoptera could have been in F. W. Hope collection; see numbered catalogue of localities found amongst his manuscripts.

BACOT, A.   Specimens used for Mendelian breeding experiments on *Acidalia virgularia* with L. B. Prout (1909).
Material on which his papers on heredity were based (1912).

BADCOCK, H. D.   Collection of British spiders, together with Arachnida, Myriapoda, Crustacea, and insects both British and foreign (presented by his sister Miss E. C. Badcock, 1939).

BADCOCK, MISS   *Cerambyx* sp. from New Holland taken alive in the Oxford Botanic Garden (1864).

BADEN, F. and SOMMER, M. C.   Collection of Coleoptera of the world sold by Mr van der Poll through Janson for *c.*£100 (presented by E. B. Poulton, 1911): Curculionidae, Brenthidae, Anthribidae, and Bruchidae (16 000 specimens, *c.*5000 species, said to have been chiefly named by Schönherr and including many of his Types, also many species determined by Faust); Carabidae (9567 specimens, *c.*3700 species, many named by Bates and Putzeys); Cerambycidae (9388 specimens, *c.*3300 species, many determined by Bates); Staphylinidae (2216 specimens, 730 species) and Silphidae (1533 specimens, 500 species); also Cassidae, Hydrophilidae, Clavicornia, Eucnemidae, Elateridae, Malacodermidae, Mordellidae, Chrysomelidae, Erotylidae, Coccinellidae, and six boxes of Phytophaga.

BAGNALL, R. S.   British Chilopoda, Symphyla, Pauropoda, Diplopoda, Thysanoptera, Isopoda, Thysanura, Collembola, Anoplura, Mallophaga, also foreign Thysanoptera (1912).
Various small donations of British insects.

BAIRSTOW, S. D.   Coleoptera from Cape of Good Hope (1882).

Collection of South African shells obtained, with few exceptions, from Algoa Bay, presented on condition 'no specimens be ever extracted excepting in exchange for other S.A. shells and that it be termed the Bairstow Collection' (1891). Specimens were verified by G. B. Sowerby and include some of his Types.

Baker, G. T. Bethune-   *see* Bethune-Baker, G. T.

Baker, J. R. and Percy Sladen Memorial Fund   Collection of insects, mainly Coleoptera, made in the New Hebrides and adjacent groups (1923, 1927).
Identifications for 1923 collection in Archives.

Balfour, C.   Over 300 insects and five Arachnids from Java collected 1871–88 (presented by Miss Balfour, 1914).

Baly, [J. S.]   Aculeate Hymenoptera purchased from Baly *via* J. O. Westwood (1857).
From auction: lot 268, Thynnidae, Tiphiidae, *Leucospis*, etc., 8*s.*; lots 255 and 256, exotic Apidae and Chalcidoidea, 3*s.*; lot 270, Mutillidae, 4*s.* (March 1877).
See also Stevens, J. C. and S. (1858).

Bamfill, —   Two bottles of snakes, etc., from Lagos (December 1870, 5*s.*).

Bang-Haas, A.   *see* Staudinger, O.

Banks, H. H.   Hymenoptera and Coleoptera from Federated Malay States (1910, 1911).

Barker, C. N.   Moths from Natal, chiefly in the neighbourhood of Durban, 1887–96 (presented through R. Trimen, 1899).
Insects, chiefly Coleoptera from the same locality, also Chimoio, Mozambique, and Grahamstown, 1889–1904 (1903, 1904, 1907).

Barlee, G.   Collection of British shells and two manuscript catalogues (1862).

Barnes, Mrs   Hymenoptera, chiefly Oriental (1912, £2).

Barnett, Dr   Insects, mainly butterflies, collected in north-west Peru, April 1931 (n.d.).

Barthélemy, L. A.   Insects and Crustacea to Hope (purchased *c.*1835–52).

Bartlett, E.   Lepidoptera from Madagascar (1878, £1. 1*s.*; 1881, 10*s.*).

Barton, E. S.   31 specimens of the rare beetle *Crioceris lilii* from Chobham and specimens illustrating its life history (1942).

Bassi, C.   66 beetles to Hope (May 1835); insects to Westwood (June 1835).

Bate, Miss D. M. A.   30 insects from Cyprus (1901).
848 insects of many orders from Cyprus (1903).
Butterflies from Crete (1904).
Set of insects from Cyprus, chiefly Nicosia (1908).

Bateman, —   Coleoptera and Hymenoptera selected from his collection made at Melbourne, Australia, *via* J. O. Westwood (1857).

Bates, H. W.   Hymenoptera etc. from Para, Brazil (purchased by Westwood, 1848).
Megacephala (Cicindelidae) from the Amazon (purchased by Hope, June 1858).
50 species of Erycinidae from the Amazon (Westwood, August 1859).
Diurnal Lepidoptera from the Amazon (February 1860, £23. 7*s.*).
Selection of Coleoptera and 473 other insects from the Amazon, selected from his private collection (Hope, 1861, £4. 11*s.* 6*d.* and £28).
314 Hemiptera and 24 termites from the Amazon (1862, £4. 16*s.* 6*d.*).
Nine Neuroptera and eight spiders from the Amazon (purchased 1863).
Hymenoptera from the Amazon region.
Bates types of Orthoptera *via* W. W. Saunders (1873).

Butterflies from Brazil in Godman–Salvin collection (1896).
See also ROTHNEY, G. A. J. (1910), *and* BADEN, F. and SOMMER, H. C. (1911).

BAZELEY, E. H.   Insects from Matatiele, Cape Colony (1906).

BAZETT, MRS E. C.   Insects from Switzerland and Spanish Pyrenees (1897, 1899).
184 insects, mainly Lepidoptera, collected at or near Mengo, Uganda (1900).
Collection of nearly 7000 British Micro-Lepidoptera (1904).

BEADLE, H. A.   Set of Lepidoptera from the North of England collected 1900–4 (1906).

BECCARI, O.   *see* HANITSCH, K. R.

BECHER, LIEUT.-COL. L. E.   151 Lepidoptera, mainly butterflies, from the Himalayas, Burma, Ceylon, and Celebes (1928). Collection given to his brother 30 years previously and believed to have been captured by his cousin Wilfred Walker.

BEDDARD, F.   Parasites of birds from Zoological Society (October 1884).

BEHRENS, CAPT. T. T., Royal Engineers   Collection of Lepidoptera from the neighbourhood of Victoria Nyanza collected 1902–3 (1906).

BELL, G. V.   137 insects of various orders and five other arthropods from the neighbourhood of Kuantan, Pahang, Federated Malay States (1913).
See also BELL, REVD J. W. B.

BELL, REVD J. W. B.   Two species of rare Buprestidae from Champion Bay, Australia (1889).
Magnificent specimen of *Vanessa antiopa* captured in the Rectory Gardens, Pyrton, Oxon (1901).
Lepidoptera comprising British specimens collected within five miles of Pyrton Rectory, 1890–1915, and specimens from world-wide localities including exotic species collected by G. V. Bell in the Federated Malay States (presented by Mrs Bell, 1920).

BELL, T.   Crustacea (52 new drawers) purchased and presented by J. O. Westwood (1862).
Reptiles, tortoises, crocodiles, and lizards, dry and in spirit, purchased and presented by F. W. Hope (1862).
The collections were valued by S. Stevens as follows: Crustacea (about 2000 specimens of 500 species) £200, reptiles (about 1065 bottles) £130, tortoises (250 specimens) £90, crocodiles and lizards, dry (about 40) £20, the rhinoceros skins and horns, which were not purchased, £25.
Some miscellaneous insects in spirit collection.

BELL MARLEY, H. W.   Collection of South African Lepidoptera (1939).
120 butterflies, 215 moths (mainly Sphingidae), and 85 Neuroptera from South Africa (1942).

BELLAMY, F. A.   Insects of many orders from Tenerife, including Hymenoptera Aculeata named by E. Saunders, also a few spiders and egg-cases, Myriapoda, and Amphipoda (1902).

BELSON, REVD W. E.   200 preserved birds, partly British and partly South American (1891).

BELT, T.   38 diurnal Lepidoptera and seven *Heterochroa* (1870).
Butterflies from Nicaragua in Godman–Salvin Collection (1896).
Two historic specimens of Orthoptera mentioned in his book *A naturalist in Nicaragua*, 1874, one of them fragmentary, presented by his son A. Belt (1948).

BENNETT, E. N.   Collection of insects made in Sokotra including two new species of

Rhopalocera, one of Orthoptera and four of Arachnida (1897). (See *Proc. Zool. Soc. Lond.* **1898,** p. 172.)

Insects of many orders captured in Egypt during the Omdurman Campaign, recorded in his book on the campaign, also a few from Palestine and Syria (1898).

Insects from South Africa and Norway (1900–2), Lofoten Islands (1903).

BESSER, W.   Coleoptera from Russian localities to Hope (1830).

BETHUNE-BAKER, G. T.   Ethiopian butterflies (1915–22).

BEWICKE, MR   *see* ORMEROD, MISS E. A. (1860).

BIGOT, J. M. F.   Exotic Asilid flies presented by G. H. Verrall (1908).

Diptera (except Tabanidae and Syrphidae which went to British Museum (Natural History)) *via* G. H. Verrall, and J. E. Collin (1967).

BILLBERG, COUNT G. J.   Cabinet *via* Hope Deed of Gift (1849). (Billberg's first collection was burnt in 1822. He formed a second collection which he sold to J. G. Children: see letter from C. O. Waterhouse to Lord Walsingham, 11 May 1897, in British Museum (Natural History).) A few specimens labelled 'Blbg' have been recognized.

BINGHAM, C. T.   Oriental Diptera and Hymenoptera, chiefly from Burma (1902, 1906, 1908).

See also ROTHNEY, G. A. J.

BIRCH, F.   156 butterflies, 12 moths, and two Odonata from San Jacintho Valley, Minas Gerais, 1907–8 (1911).

BIRD, REVD C. S.   A few insects to Hope (1830–1).

BLACHIER, C.   50 butterflies from Chandolin, Switzerland (1899).

BLACKWALL, J.   Remains of his spider collection *via* O. Pickard-Cambridge (1917).

BLAKE, J.   *see* MILLAR, H. M. and BLAKE, J.

BLANKENHORN, DR   Collection of microscope slides for exhibiting the *Phylloxera, Coccus vitis, Tyroglyphus*, etc., and other insects (J. S. Madden, 1881, £5. 5s.).

BLASCHKA, L. (of Dresden)   Forty models of sea anemones; price with postage and carriage through Messrs Parker £4. 7s. 2d. (1867). Now on display in University Museum.

BLASDALE, P.   Series of African Asilidae (1955).

BLOOD, B. N.   About 2500 microscope slides of Hymenoptera Chalcidoidea, rich in the families Mymaridae and Trichogrammatidae, including some Types (Mrs Blood, 1951, £10).

BOHEMAN, C. H.   Coleoptera from Natal (1860).

Some Coleoptera Types in Baden–Sommer Collection.

BOLIVAR, I.   Spanish Orthoptera and insects of various orders from tropical West Africa (1902).

BOND, F.   Two specimens of Serotine bats from Freshwater, Isle of Wight (1861).

25 fossil crabs from Cambridgeshire (1864).

BONDAR, G.   Hymenoptera from south-east Brazil (presented by G. A. J. Rothney, 1917).

BONVOULOIR, MARQUIS H. DE   Collection of beetles from Madrid, La Granya, Escurial, Hautes Pyrénées, and Perpignan, presented to J. O. Westwood and thence to collections (1866).

BORMANS, A. DE   Collection of Orthoptera including many Types (£9. 14s., including some monographs, 1900).

Bostock, E. D.   Lepidoptera from various British localities (1914).
Some moths from Cambridge district (1923).

Boucard, A. (Natural History Agent)   Insects of all orders from Mexico (April 1868, Fr.72.50 = £2. 18s., and September 1868, 4s. 2d.).
Eight Coleoptera from New Caledonia and Uruguay (February 1871, 10s.).
Two Lepidoptera, five Arachnida, and three Homoptera (July 1871, 10s. 6d.).
Selected Coleoptera (July 1871, £4. 1s. 6d.).
A few insects from Natal, New Holland, Australia, New Caledonia, California, Borneo, Cayenne, Mexico, and Brazil (June 1872, 16s.).
Various insects (July 1886, £6. 15s. 8d., and May 1889, £8. 6s. and 3s. 10d.).

Boulay, F. H. du   404 insects, mainly Coleoptera, from the Swan River (January 1869, £10).
Various beetles from Australia (purchased from Mr Cutter, November 1869).

Bourke, Rear Admiral E. G.   Lepidoptera collected in various parts of the tropics, arranged geographically in 19 cabinets (1923). The collection was commenced in 1865 when he was on the west coast of Africa for the suppression of the slave trade. He directed the collection should remain intact, although he had no objection to particular specimens being removed to the General Collection when their removal would add to interest and promote their study, provided they were labelled as part of the Bourke Collection. Kept separate.

Bowell, E. W.   Two specimens of the rare beetle *Buprestis chryseis* (synonym of *Agrilus sinuatus*) from the New Forest (1891).

Bowes, A. J. L.   99 bred specimens of British Ichneumonidae and Braconidae (1939).

Bowring, C. T.   219 butterflies and 18 moths from Wenchow district of south-east China (presented by J. J. Walker, 1913).
A few butterflies from Hainan Island, south China (1919).

Boyd, W. C.   Collection of British Lepidoptera (1901).

Boys, Capt. W. J. E.   Insects to Hope (1842).
Indian Hymenoptera, Diptera, Hemiptera, etc., with cabinet, purchased in 1848 by J. O. Westwood (1857).

Boys' Home, Tananarive   Madagascar butterflies (1920, £3. 3s.).

Brancsik, K.   59 Orthoptera, including Types of many species described by donor (1907).

Brauer, F. M.   Series of European species of Oestridae and larvae in spirit; also Neuroptera [sens. lat.] of the genera *Palingenia*, *Chrysopa*, *Osmylus*, *Ascalaphus*, *Panorpa*, *Bittacus*, and *Theleproctophylla* (presented by J. O. Westwood, 1863).

Bree, Revd W.   Nine specimens of *Apatura iris* and *Polyommatus arion* (1860).

Brême, Marquis F. de   Insects to Hope (1841–2).

Brewer, [J. A.]   Purchased at sale of E. Brown collection, included in lot 345, 52 Carabidae from Cartagena, Spain (March 1877).

British Expedition to East Greenland, 1935–6   *see* Wager, L. R.

British Museum (Natural History)   Specimens by donation and exchange over many years.

British South African Company   *see* Neave, S. A. (1906).

Britten, H.   268 Psocidae, representing nearly 40 species, from Oxfordshire (1918).
12 Trichopterygidae (Coleoptera) collected on Seychelles Expedition, including paratypes of six species described by donor (1926).
191 Hymenoptera Parasitica (presented by M. F. Claridge, 1958).

BROADHEAD, E.   38 slides of *Liposcelis* (Psocoptera) including paratypes (1947).
15 slides, Types and paratypes, of Psocoptera described by donor (1951).
36 slides of Psocoptera, including some paratypes (1952) and a series of macropterous and brachypterous *Psoquilla marginepunctata* (1953).
Seven slides *Liposcelis albothoracicus* and *Lepolepis bicolor*, and four of *Liposcelis obscurus*, holotypes and paratypes (1955).

BRODIE, A. J.   50 insects and a spider's egg-cocoon from Epe, Lagos, West Africa (1900).

BRODIE, N. S.   World collection of Lepidoptera. Approximately 30 000 butterflies and many thousands of moths in 1000 store boxes, chiefly from India, Penang, and North Borneo, also Africa, Highlands of Central Asia, circumpolar regions, Malay Archipelago, and tropical South America (presented by Mrs Brodie, 1933).

BROWN, C. J.   Butterflies collected in Ranikhet and Lucknow, 1923 (presented by R. H. Rattray, 1932).

BROWN, E.   Over 1500 specimens of Coleoptera and Hymenoptera Formicidae (purchased at auction March 1877, £7. 11*s*. 6*d*.) (Includes some Carabidae from Wallace and Brewer.)

BROWN, E. S.   23 Diptera, comprising six species, from Hertford Heath (1940).
21 aquatic Hemiptera from Suffolk (1942).
73 Corixidae and four Gerridae from Derbyshire and the Faröe Islands (1946).
British insects of the orders Dermaptera, Diptera (Syrphidae and Tipulidae), Coleoptera (mainly aquatic), Hemiptera (mainly aquatic), Odonata, Orthoptera; also aquatic Coleoptera and Hemiptera, and miscellaneous Lepidoptera and Orthoptera from various localities in the Middle East, East Africa, and Australasia (1972).

BROWN, F. N.   Extensive collection of Hymenoptera Aculeata from Orange River Colony and Natal and various insects from Natal and Mashonaland (1902); Hymenoptera Aculeata from South Africa (1903).

BROWN, H. ROWLAND-   *see* ROWLAND-BROWN, H.

BRUCK, E. VOM   Coleoptera sent to J. O. Westwood (1848), many determined by Erichson, Kiesenwetter, and Märkel; specimens with red labels originals of Erichson, green those of Kiesenwetter.

BRUNNER VON WATTENWYL, C.   *see* BURR, M. (1903).

BRYDGES, —   Insects from Chili to Hope (*c*.1834).

BUCKLER, C. A.   A specimen of the Bohemian waxwing bird (1863).

BUCKLEY, [C].   Insects, mainly Coleoptera and Lepidoptera, from Ecuador (Mr Higgins, April 1872, £4. 15*s*.).

BULLOCK, G. H.   Series of Lepidoptera, chiefly butterflies, from Fernando Po, Peru, and the 1921 Everest Expedition 'with the permission of the Mount Everest Committee' (1921).
See also DONCASTER, L. (1927).

BURCHELL, W. J.   '1865. The whole of the *Burchellian Collections* of Natural History (except the Plants which were refused by the University & given to the Kew Museum) were presented by Miss Burchell . . . The Birds, Reptiles, Skulls &c. have been distributed to the various departments of the Museum . . .' (extract from Westwood's journal).
'The only condition that I should wish to annex to the gift is that the Collections should be distinguished as those of my late Brother' (extract from letter from Anna Burchell dated 8 April 1865).

These historic collections comprise material from South Africa, 1810–15, Portugal and Tenerife, 1825, and eastern Brazil, 1825–30. All specimens bear a distinctive label. The collections also contain British insects, some specimens from St Helena, China (not captured by Burchell), and a series from Van Diemen's Land captured by his sister Harriet Butcher (q.v.).

See also Archives.

The following list of African and Brazilian material has been extracted from a letter from W. Goddard Jackson to the Trustees of the New Museum, Oxford, 20 August 1864, offering the collections on behalf of Miss Anna Burchell (Oxford University Archives, UM/M/1/Z, f. 61^r):

The Collections consist of first those from Africa (viz.) of

Quadrupeds	120 skins	95	specimens [*recte* species]	
Birds	541	of 268	,,	,,
Reptiles	70			
Insects	2315	of 815	,,	,,
Plantae	60,000	5,000	,,	,,
Minerals	110			
Serpentes	50	27	,,	,,
The Brazilian				
Quadrupeds	64	39	,,	,,
Birds	817	362	,,	,,
Reptiles	110	76	,,	,,
Insects	16,000			
Plantae	75,000	of 7,022	,,	,,
Minerals	210			
Pisces	25	23	,,	,,
Shells	500	100	,,	,,
Radiata		11		
Annelida	3	3		

Burmeister, H. Rare Coleoptera from the Argentine (1889).

Burnett, R. 55 butterflies, mainly *Euploea* and Danaines, from Hong Kong and Tonkin (1949).

Burr, M. A few British Orthoptera from Zoological Gardens, London, and Kew Gardens (1896).

Set of Forficulidae captured by Fruhstorfer in Lombok; ten Odonata and four *Raphidia* from Black Pond, Esher (1897).

A few Orthoptera and one Forficulid from Ceylon, Java, and Celebes; Orthoptera and Lepidoptera from south-east Europe; insects of all orders from São Paulo, Brazil (1898).

Insects from Port Elizabeth, South Africa and many species of Orthoptera from Montenegro, south-east Europe, Ceylon, North and South America, and Madagascar (1900).

Exotic Orthoptera, including specimens from Rio Grande do Sul received from Brunner von Wattenwyl (1903).

Collection of Palaearctic Orthoptera, containing many Types (1922, £50). Specimens rearranged and checked by Dr Uvarov.

64 Orthoptera, part of a collection described by J. A. G. Rehn, collected in the Rio district of Brazil (1925).

Burras, A. E. Lepidoptera from the Basses Alpes and Pyrenees (1925) and Hautes Alpes (1926).

Burtt, E. and B. D.   1669 specimens of British Hymenoptera (1938).
52 Diptera including original series of the rare Dipteron *Tabanus glaucopis* taken in 1922–3 by B. D. Burtt (1939).
Butterflies from Tanganyika Territory collected by B. D. Burtt (1940).
Large collection of insects of various orders from 'Flamboyant Trees' infested with Bostrychidae, including *Charaxes* at oozing juice; also many giant *Palophus* (Phasmidae) from Tanganyika Territory (1948).

Butcher, Mrs H. (née Burchell)   Series of insects collected at Lowlands, near Richmond, Van Diemen's Land, brought to England February 1836, part of Burchell Collection (presented by Miss Anna Burchell, 1865).

Butler, Revd A. G.   98 Lepidoptera from Upper Engadine and Brieg, Valais, Switzerland (1898).

Butterworth, F.   56 insects captured at Toowoomba, near Brisbane (presented by W. J. Lucas, 1904).

Buxton, P. A.   Some butterflies from New Hebrides and New Caledonia (*c.*1925).
See also Morice, Revd F. D.

Buysson, Vicomte R. du   56 wasps (Diploptera) from various localities, especially Mexico, determined by the donor (1907).
See also Morice, Revd F. D.

Byatt, H. A.   Lepidoptera from Lake Nyassa, British Central Africa (1902).
Butterflies and a few other insects collected by natives, 1898–9, in north-eastern Rhodesia (1904).
Lepidoptera, chiefly butterflies, from British Central Africa and Somaliland (1904–6).

Byers, G. W.   Series of North American Panorpidae (1962).

Cambridge University Expeditions   Material in both dried and spirit collections.
Jan Mayen 1921, 19 spiders (presented by W. S. Bristowe, 1948); Spitsbergen 1927, from Edge Island (collected by A. P. G. Michelmore); Bear Island 1932 (collected by D. Lack).

Cameron, P.   British Tenthredinidae and Cynipoidea, some possibly syntypes (1884, £10; includes cost of naming collection of British sawflies).
Cameron Types of Rothney's Oriental Hymenoptera, mainly Indian.

Cantlie, Sir K.   About 6000 butterflies of all families mainly from Assam, also 8000–9000 unset specimens in papers (1972).

Cantor, T.   Duplicates from his Chusan insects, collection from the Khasia Hills (some presented to Cantor by Mr Griffith), terrestrial and aquatic Testacea for 'the worthy Professor Kidd for the Oxford Museum' and insects from Prince of Wales Island (Púlo Pínang). Sent to F. W. Hope 1841, 1842, and 1846 (1849).

Cardew, A. G.   Lepidoptera from Nilgiri Hills and Cannanore, India (1896).
Set of moths from Cornwall (1899).

Carmichael, Lord   [Indian insects, especially Aculeate Hymenoptera (1915)].
Named insects (Rhynchota, Hymenoptera, Coleoptera, and Orthoptera) also Reptilia, Arachnida, and Myriapoda (1920).

Carpenter, G. D. Hale   Lepidoptera from Victoria Nyanza, Damba Island, Uganda, Natal, Tanganyika Territory, ex-German East Africa, and Portuguese East Africa (1911–30).
350 insects of various orders from the Rhône Delta, including the rare Acridian *Prionotropis hystrix rhodanica*, a relic of an ancient desert fauna (1938).
Many specimens of British insects donated during his tenure of the Hope Chair.

CARR, L. A.   British Braconidae (1926, £5. 10s.), sawflies (1929, £5), and Argentine spiders (1929, £2. 2s.).

CARSON, G. M.   Insects of many orders from New Guinea (1912).

CARTER, A. E. J.   *see* VERRALL, G. H. and COLLIN, J. E.

CASTELNAU, COMTE DE   *see* LAPORTE, F. L. N. DE C. DE, COMTE DE CASTELNAU.

CESNOLA, CONTE DI   129 butterflies and five Arctiid moths from Piedmont, Italy (1906).

CHABRIER FILS   Insects to Hope (1835).

CHAMPION, G. C.   5200 beetles from his collection (presented by his son H. G. Champion, 1936).
Butterflies collected by G. C. Champion in Central America in Godman–Salvin Collection (1896).

CHAMPION, H. G.   Lepidoptera from United Provinces, North India, and south-west Tibet (1915–22, 1925).
34 Continental butterflies and six British Lycaenids (1948).

CHAMPION, R. J. (youngest son of G. C. Champion).   British Lepidoptera and Hymenoptera (presented by his parents, 1919).
83 Micro-Lepidoptera from Surrey (presented by H. G. Champion, 1948).

CHAPLIN, D.   Butterflies captured near Durban, Natal (presented by F. D. Godman and O. Salvin, 1897).

CHAPMAN, T. A.   Spirit material to Westwood (*c.*1869–70).
Lepidoptera from Switzerland, East Carinthia, and Norway (1899).
Insects of various orders from Spain, also a few from France and Switzerland (1902–3).
Asilids and prey from Spain (1905).
Numerous small donations over many years.

CHAUDOIR, BARON M. DE   Insects from Russia to Hope (1836).

CHESTERTON, —   Newfoundland Lepidoptera (Mr Hewitson, July 1874, 7s. and 19s.).
Collections also contain some Orthoptera.

CHEVROLAT, L. A. A.   A few insects, chiefly Mexican, to J. O. Westwood (May 1835).
Coleoptera from Barbary to F. W. Hope (September 1842).
Large collection of Coleoptera (Carabidae) forming part of Van der Poll collection (purchased and presented by E. B. Poulton, 1909).

CHITTY, A. J.   British insects of all orders (presented by Mrs Chitty, 1908). [The labels on some of his specimens need explanation: 'Huntingfield' is the name of his house at Faversham, Kent, his 'Scotland' is Loch Awe, and 'Richmond' is in Yorkshire.]

CHOPARD, L.   *see* HANITSCH, K. R.

CLARK, BRACY   Some Diptera Types *via* J. O. Westwood, including those of *Cuterebra atrox* and *C. detrudator*, described in 1848.

CLARK, B. PRESTON   Lepidoptera Sphingidae, including several syntypes of American species (by exchange, 1915–22, 1930).

CLELAND, J. R.   Small collection of insects from South African localities and Mombasa, including a few specimens from Mozambique, the Red Sea, and Mediterranean (1905).

CLEMENT, E.   772 insects of many orders from Towranna Plains, West Australia;

rich in moths, but also includes a fine set of Australian wasps for mimicry series (1899, £5).

Cockerell, T. D. A.   North American butterflies, also interesting variety of *Pyrameis cardui* from Porto Santo (1915–22).

Codrington, Revd R. H.   A few insects from New Holland [Australia] (1864).

Coffin, E. Pine   Five double boxes containing Mexican insects of all orders *via* J. O. Westwood (1857).

Colenutt, G. W.   Fossil Diptera of genera *Gymnastes* (Tipulidae) and *Odontomyia* (Stratiomyidae) from the Oligocene freshwater formation of the Isle of Wight (1931). Exhibited at Royal Society Soirée, May 1931, by E. B. Poulton.

Collin, J. E.   *see* Verrall, G. H. and Collin, J. E.

Collins, J. J.   Insects from Oxford district and a series of Diptera (1925, 1935).
    British Hemiptera and Hymenoptera (1941, £6).

Colonial Office, Entomological Research Committee   Insects of all orders collected by their travelling entomologists, S. A. Neave and J. J. Simpson and others in Nyasaland, German and British East Africa, and West Africa (1910, 1911, 1913).

Cook, A. L.   700 butterflies from Sandakan, North Borneo (presented by H. Druce, 1899).

Cooke, Brig. B. H. H.   Collection of European Lepidoptera, comprising 12 385 butterflies and 24 drawers of moths (1946); Professor Carpenter undertook to keep this collection in its original arrangement as far as possible.

Cooke, J. A. L.   42 British spiders (1962).
    52 spiders from British and foreign localities, including Type material (1967).
    183 bottles of Arachnida, mainly spiders from many parts of the world (1971).
    See also Lampel, G. P. (1962).

Corporaal, [J. B.]   55 Lycaenidae and Hesperiidae from Billiton, Indonesia (presented by E. J. Nieuwenhuis, 1971).

Costa, A. and O. G.   Insects to Hope and possibly Crustacea (*c.*1834–55).

Cott, H. B.   96 insects from Bamburgh, Northumberland and 176 specimens of *Scatophaga* with prey (1936).

Cowan, Revd W. D.   Collection of insects from Madagascar (October 1882, £8).
    Set of butterflies collected or bred in Bétsiléo, Madagascar (1900).

Cox, Mrs E. (artist-naturalist)   Insects from Tasmania (1896).

Cox, H. E.   Collection of Coleoptera, including the British collection on which his book *A handbook of the Coleoptera or beetles, of Great Britain and Ireland*, 1874, was based, and a special collection of Heteromera, the latter containing Heteromera from W. W. Saunders (presented by Mrs Cox, 1915, 1922) (kept separate). Many books also given to Library.

Crabtree, B. H.   Lepidoptera from various British localities (1914, 1923, 1925).

Craig, Mrs E. S.   Nine insects and spiders from Estância do Rey, Mercedes, Uruguay (1902) and 14 Lepidoptera (1913).

Craig, R. A.   Lepidoptera from Stonecutters Island, Hong Kong (1912).

Crawford, L.   Stems of Eucalypts from Australia, infested with Coccidae (1889).

Crawford, O. G. S.   Small collection of insects from Blue Nile (1915–22).

Crawford, W. M.   Butterflies from Orissa, north-east India (1915–22).

CRAWLEY, W. C.   Ants of the world, containing many syntypes of species described by Forel, and specimens received by gift and exchange from Emery, Santschi, Menozzi, Finzi, and other eminent myrmecologists, also many books (1929).

CRAWSHAY, REVD G. A.   British Coleoptera, including local and rare species (1905).

CREMER, J. A.   35 Lepidoptera and one fine Mutillid from Lagos (1906).

CRÉMIÈRE, —   Insects to Hope (1842).

CROMPTON, E. and F.   Insects captured in neighbourhood of Victoria, Vancouver Island (1897).

CROSS, REVD L. B.   Insects of various orders taken during voyage to and from Australia; includes specimens from Sydney, New Zealand, Ceylon, Java, Penang, Curaçao, and Panama (1934).
    100 hornets from a nest in a tree at Tubney, preserved in spirit (1938).

CROSSE, W. H.   Large collection of insects from tropical West Africa (1902).

CRUDGINGTON, I. M.   65 butterflies from Cape Town (1946).

CRUDGINGTON, I. M. and DICKSON, C. G. C.   Rare butterflies peculiar to Cape Province, South Africa (1946).

CUMING, H.   Insects to Hope (*c*.1841).
    Nest of gigantic caterpillar (*Cossus?*) (1863).
    A number of Cirripeds (from sale by auction of his collection 1866, 6*s*. 6*d*.).
    Five bottles of Crustacea in spirits (Mr Geale, London, 1866, 5*s*.).

CURTEIS, —   [Curties, T.]   85 insects from Swan River, Australia (1869, £2. 10*s*.).

CURTIS, C. M.   A few insects from Madeira to Hope (1836).

CURTIS, J.   *Dielocerus ellisii* ♀, honeybee from New South Wales (by exchange, 1848).
    Series of galls of *Cecidoses eremita* with the pupa, and galls of three other species from his collection (received from his widow and presented by J. O. Westwood, 1863).
    A large number of Hymenoptera purchased at sale of Entomological Society's British Collection (July 1863, lot 125, 2*s*. 6*d*.).
    11 exotic Hymenoptera and Diptera from Curtis Collection (S. Stevens, November 1865, 8*s*. 3*d*.).
    A few specimens of British Chalcidoidea, probably syntypes, collected by him.
    The Curtis Collection was offered to Oxford in 1858, but not accepted.
    See DALE, J. C. and C. W., *and also* IMAGE, S.

CUTTER, W.   One *Papilio* and three Nymphalidae from Old Calabar 8*s*., and various beetles from same locality 6*s*. (1866).
    Coleoptera and one spider (1867).
    Moloch horridus 7*s*. (1868).
    Various insects 13*s*., three butterflies from Bogotà 6*s*., insects from Port Natal 8*s*. 6*d*., and one butterfly from Port Natal 8*s*. (1868).
    Insects from Japan, Australia, and Cameroons £2. 10*s*., two small Lucanidae from Burma 1*s*. (1869).
    One *Sphinx* 1*s*., various insects £1. 14*s*. 6*d*., and four Paussidae 14*s*. (1870).
    Two *Cerapterus* from Queensland, 8*s*. (1872).
    One moth and one butterfly 4*s*. 6*d*., and 31 butterflies from Old Calabar £3. 15*s*. (1874).
    One *Meloe* and one *Cetonia* from Madagascar 3*s*. 6*d*. (1875).

DADD, E. M.   European and North American-bred Catocalid moths (1912).

DAHLBOM, A. G.   Insects from Museum of University of Lund (1858).

DALBY, REVD A. Lepidoptera from Java and Sumatra (1939).
Butterflies collected on a tour of the East Indies (1941).

DALE, J. C. and C. W. Collection mainly of British insects of unique historic interest begun by J. C. Dale. Bequeathed with additions by his son C. W. Dale (a great friend of J. O. Westwood) in 1906 on condition that it should be kept permanently separate under the name of the 'Dale Collection'. Originally comprised seven cabinets of British Lepidoptera, three cabinets of foreign Lepidoptera, five cabinets of British Coleoptera, 14 cabinets of British Hymenoptera and Diptera, four cabinets of Odonata, seven or eight cabinets of shells, eggs, etc. The Odonata and Orthoptera have been transferred to modern cabinets in the general collection for safety.

The collection contains some Types of Lepidoptera, described by J. Curtis, collected during Sir John Ross's second voyage in search of a north-west passage, 1829–33, a few Selys Odonata Types, some syntypes of Walker Chalcidoidea, a number of Haliday specimens, insects collected in France by Curtis in 1830, and some taken by F. Walker in Sweden and Norway in 1836. There are also four drawers of Wollaston beetles from Madeira, Cape Verde, Canary Islands, and St Helena and five drawers of shells from the Atlantic Islands.

Two bronze celts which had formed part of the Dale Collection were transferred to the Ashmolean Museum in 1961.

DALTRY, H. W. British insects of all orders and notebooks given to Dr M. W. R. de V. Graham, and placed on loan to the Hope Department, were donated to the Herbert Art Gallery and Museum, Coventry in 1983.

DALTRY, REVD S. J. Collection of Lepidoptera made in the neighbourhood of Urumiah, north-west Persia (presented by R. W. T. Günther, 1898).

DAMELL, E. 96 insects from Fiji Islands and various parts of Australia (May 1863, £4).
Various insects from Australia (purchased from S. Stevens, July 1863).
Coleoptera collected by Damell (purchased from Higgins, 1866).
22 Coleoptera and ♂ and ♀ *Saturnia eupestris* (1869, 10 thalers = £1. 10*s*.).

DAMMERS, CDR. C. M. Butterflies from Concepción, Province of Tucuman, north-west Argentina (presented by J. Blamey, 1912, 1913).
A few butterflies from various localities in America (1931) and central California (1934).
Eight specimens of *Megathymus* (Hesperioidea) from California (1935).

DARBISHIRE, C. J. Coleoptera, mainly water-beetles, to Hope (1830).

DARLINGTON, P. J. 30 specimens of Hemiptera and Coleoptera from Cuba, illustrative of experiments on mimicry (1939).

DARWIN, C. Many specimens *via* F. W. Hope are scattered through the old collections, and so far the following have come to light:

			Number of Specimens
Coleoptera			
Lucanidae	*Dorcus darwinii* Hope	Chili	1
Carabidae	*Carabus darwinii* Hope	Chiloë	1
	Carabus suturalis Fabr.	Tierra del Fuego	2
	Carabus chiloensis Esch.	Chiloë	1
	Carabus insularis Hope	Chiloë	3
Dytiscidae	*Cybister biungulatus* Bab.	Maldonado, Rio del Plata	1
	?'Hydroporus' (un-named)	Sydney	1

Hemiptera—Heteroptera			
Pentatomidae	*Canthecona* sp.	Sydney	1
	near *Nezara*	Sydney	4
	near *Nezara*	King George's Sound	1
	near *Nezara*	Hobart	1
	Dinocoris sp.	Sydney and Hobart	3
	Elasmostethus sp.	Sydney	1
Coreidae	*Amorbus* sp.	Hobart	1
Lygaeidae	*Graptostethus* sp.	Sydney	2
Reduviidae	immature	Sydney	3
Corixidae	*Sigara australis* Fieb. (det. by Kirkaldy)	Sydney	5

Hemiptera—Homoptera			
Cicadidae	*Melampsalta* sp.	Sydney	1
Flatidae	*Carthaea* sp.	Sydney	2
Cercopidae	*Orthoraphia* sp. [*Bathyllus*]	Sydney and Hobart	8
Unidentified genera		Sydney and Hobart	6

Hymenoptera Chalcidoidea			
Chalcididae		Sydney	1

Siphonaptera			
Pulicidae	*Pulex irritans* Linn. (Denny Coll.)	Chiloë	1

Mallophaga and Anoplura
  Some from South America in Denny Coll.

Microlepidoptera		Sydney, Hobart, and Tahiti	9

Neuroptera			
Mantispidae	*Mantispa* sp.	Hobart	1

Orthoptera			
Acrididae		Sydney and Hobart	8

Diptera			
Asilidae	*Bathypogon* sp.	Tasmania	1

Flies collected in Australia and Tasmania were exhibited at a soirée in the Examination Schools, Oxford, in February 1909.

Crustacea *via* Bell Collection (1862)

A work entitled 'Darwin's insects' by K. G. V. Smith dealing with specimens from the *Beagle* voyage will be published by the British Museum (Natural History) in their *Bulletin* (Historical Series).

DAWKINS, [W. B.]   Three fossil Crustacea from clay near Cambridge and cast of Egyptian scarab in British Museum (1862).

DAY, C. D.   324 microscope slides of British Hymenoptera Chalcidoidea, which have been incorporated with the Blood Collection, and collection of 841 British Homoptera (1954).
  410 British Diptera (Tachinidae and Calliphoridae) and Hymenoptera (1963).

DE BORMANS, A.   *see* BORMANS, A. DE.

DE HAAN, W.   *see* HAAN, W. DE.

DEJEAN, COMTE P. F. M. A.   Around 50 Coleoptera sent to Hope in 1823 (see letters from Dejean to Hope, 26 October 1822 and 21 June 1823).
  Some Lepidoptera out of Marchal Collection. (According to Horn, W. and Kahle, I. (1935–7). Über entomologische Sammlungen, Entomologen & Entomo-Museologie. *Ent. Beih. Berl.-Dahlem* **2–4,** i–vi, 1–536.)
  See also LATREILLE, P. A.

De Laporte, F. L. N. de C., Comte de Castelnau   *see* Laporte, F. L. N. de C. de, Comte
de Castelnau.

Denny, H.   Collection of Anoplura Exotica and Britannica with illustrations and
manuscript descriptions (purchased by J. O. Westwood from the representatives
of the late H. Denny, 1871, £30). Includes some Darwin material from South
America. Kept separate.

Denton, Sir G.   Lepidoptera Rhopalocera from Gambia, West Africa (1903).

Denton, Mrs J. M. A.   Scorpion and a cicada from Ilaro, near Lagos, Nigeria (1899).
Collection of Lepidoptera made in the Lagos district (presented by Sir George
Denton, 1919).

Depuiset, A.   Insects from world-wide localities (F. W. Hope, September 1860, Fr.
280).
Lepidoptera (1862, Fr. 100 = £4).
Lepidoptera, Coleoptera, etc. (August 1865, Fr. 87 = £3. 10*s*.).
Coleoptera collected in Chili by M. Germain and other insects from the Moluccas
(£2), insects and spiders (1866, Fr. 62 = £2. 10*s*.).
Lepidoptera (1868, Fr. 111 = £4. 9*s*. 2*d*.).

Desjardins, J. F.   Specimens from Mauritius sent to J. O. Westwood (April 1835).

Desvignes, T.   British Ichneumonidae named by Desvignes *via* J. O. Westwood
(1857).

Deutsches Entomologisches National-Museum Berlin   Blattidae, including a Type
and some syntypes *via* R. Shelford (1907).

Deyrolle, A., H. and E.   Mexican Coleoptera and a few from Chili (purchased by
F. W. Hope, 1860, Fr. 13.70).
Coleoptera (1863, Fr. 21.75 = 18*s*. 2*d*.).
Insects chiefly from Chili (1866, Fr. 185.90 = £7. 9*s*.).
Lepidoptera from Cayenne through Mr Hewitson (1867, 32 specimens at 1*s*. 3*d*.
each, £2).
Insects and Madagascar reptiles (January 1885, £5. 1*s*.).
Extensive series of insects, mainly Coleoptera, from Morocco (October 1885,
£8. 2*s*. 2*d*.).
Calognathus mulleri and other Coleoptera (1886, £10).
Exotic Coleoptera including *Hypocephalus armatus* ♂, from the interior of Brazil
(1891, £15. 6*s*.).
Various unspecified insects were also purchased between 1865 and 1890.

Dickson, C. G. C.   Butterflies from Zululand and Cape Province (1946).
14 Lycaenidae, one *Charaxes pelias pelias*, and two *Phasis dicksoni* from Cape
Province (1948).
11 butterflies from Cape Province (1949).
70 butterflies, mainly Lycaenidae and a few Satyridae and Hesperiidae, from
Cape Province (1971).
136 Lycaenidae from South Africa (1972).
93 Lycaenidae from Cape Province (1973).
16 Lycaenidae from Cape Town area (1975).
See also Crudgington, I. M. and Dickson, C. G. C. (1946).

Dillenius, J. J.   Cabinet *via* Hope Deed of Gift (1849).

Dixey, F. A.   Insects from Mortehoe, North Devon, and five *Araschnia prorsa-levana*
bred from European pupae (1897).
Butterflies from Sarawak (1898).

105 British moths (1903).

Specimens from South Africa (1905).

Insects taken on the voyage from Australia in 1914, and in 1922 at Lisbon, Madeira, Tenerife, and Grand Canary (1915–22).

Donated many other British and foreign specimens over the years whilst associated with the department, also a large collection of microscope slides.

DOBIASCH, E.   Insects from Croatia (October 1888, £1. 11s.).

DODD, F. P.   Specimens illustrating *Notes upon some remarkable parasitic insects from North Queensland* (*Trans. ent. Soc. Lond.* **1906,** 119–24), including Types described by C. T. Bingham (1906).

Collection of insects chiefly Hymenoptera and a few Arachnids from Queensland (presented by G. A. J. Rothney, 1911).

North Australian insects of various groups (1915–22).

DODSON, MISS A. H.   135 butterflies and 32 other insects from Iran (1948).

150 butterflies, 17 moths, and five other insects from Iran (1949).

DOHERTY, W.   *see* ROSENBERG, W. F. (1902) *and also* DONCASTER, L. (1927).

DOHRN, A.   Insects from Pyrenees, Carniola, Hungary, Turkey, Abyssinia, Natal, and Philippines (by exchange 1861, 1862).

Carabus procerubes and *Damaster fortunei* (with elytra gaping at tip) from Japan (1865).

Large quantity of Invertebrata (especially 'Articulata'), in spirit, from the Mediterranean (April 1874, £10).

DOLLMAN, H.   Butterflies from Rhodesia, collected 1908 (1909).

Series of 15 Rhodesian Staphylinid beetles, named by M. Cameron, including some paratypes (presented by H. St J. K. Donisthorpe, 1929).

DONCASTER, L.   Extensive private collection of moths, from world-wide localities, the species determined by comparison with series in the British Museum (Natural History) (purchased 1927, £74. 17s. 6d.). Contains material collected by: G. H. Bullock (Abyssinia), W. Doherty (India, Java, New Guinea), Capt. J. Fraser (Uganda), H. Fruhstorfer (Java), C. Hofer (Austria), H. A. Junod (West Africa), Kricheldorf Collection (India), J. H. Leech (India), Dr Lidderdale (India), H. McArthur (India), J. G. Pilcher (Sikkim, India), A. E. Pratt (China), H. J. S. Pryer (Japan), S. Robson (India), Capt. G. Young (India), and others.

See also WATKINS and DONCASTER.

DONISTHORPE, H. ST J. K.   Large donations of British insects, mainly Coleoptera (1899–1924).

Collection of ants and other insects associated with them purchased for £100 in 1927 and retained temporarily by Donisthorpe, but resold by him in 1933 to British Museum (Natural History) and costs incurred then refunded to Hope Department.

Collection of British Coleoptera including a remarkable series from Windsor Forest (1927) with further additions (1928–43). Kept separate.

DONOVAN, LIEUT.-COL. C.   Butterflies from various Indian localities (1915–22).

Moths from the Faringdon district for British Collection (1925).

Bred British Lepidoptera (1926).

DOS PASSOS, C. F.   19 species of rare or local Papilios from North America (1947).

DOUBLEDAY, B. S.   Approximately 500 microscope slides of British Chalcidoidea, many bred from known hosts, much unmounted material, and a number of books and offprints on Hymenoptera (1956).

DOUBLEDAY, E.   A few Hymenoptera from his North American tour in 1837.

DOUBLEDAY, H.　167 Lepidoptera (August 1863).
　100 European Micro-Lepidoptera (presented by J. Collins, 1910).

DOWDESWELL, W. H.　59 butterflies and 115 moths from the island of Cara, Argyllshire (1935).
　37 moths, chiefly Noctuidae, and 42 Hymenoptera, mainly *Bombus* spp., from Cara (1936).
　22 butterflies from Natal and two from Réunion (1948).

DOWDESWELL, W. H. and FORD, E. B.　144 Bombidae from Scilly Islands (1939).

DOWNES, COL. A. M.　49 butterflies and 13 moths from Mauritius (1907).

DREWSEN, C.　25 named Danish *Vespa*, ten *Sirex juvencus*, and allied species, 15 species of Hymenoptera from various countries, *Prionus* sp. from Barbary, and various larvae (August 1858).

DRINKWATER, MRS　37 species of insects from Upper India (1886).

DRUCE, H.　Bottle of Lepidopterous and Coleopterous larvae from Chiriqui, Central America, two specimens of *Acraea rabbaei* Ward from Delagoa Bay, and one specimen of 'Callisphyris?... black species' (1886).
　Two rare locusts from Borneo (1890).
　Collection of Lepidoptera from Sierra Leone, West Africa, and South and Central America (1897).
　Moths from tropical America (1898) and Central America (1899).
　Insects of many orders from Sandakan, north Borneo, including a fine series of butterflies (1899).
　Lepidoptera and insects of other orders from world-wide localities (1900, 1902, 1906).
　Moths collected in the Khasia Hills, Assam (1907, 1908, 1910).
　20 butterflies from western China and 30 from the Ja River, Cameroons (1908).
　32 Peruvian moths named by donor (1909).
　Lepidoptera from Dutch and British New Guinea, Borneo, Philippines, Central and South America, West Africa, etc. (1910-12). Includes material collected by A. E., C. B., and F. B. Pratt, and M. G. Palmer.

DRUCE, H. H. C. J.　Insects and a few spiders from Bilderlingshof, western Russia (1902).
　25 insects of various orders from St Moritz and Ragatz (1904).
　Six Hesperiidae from various localities (1909).

DRURY, DRU　Cabinet *via* Hope Deed of Gift (1849).

DUFFIELD, J. E.　680 insects of all orders from south-western United States (1932).
　114 insects from Suez and Port Tewfik, Egypt, Colombo, Ceylon, and at sea (1942).

DUFOUR, [L.]　Two cabinets *via* Hope Deed of Gift (1849).

DURRANT, J. H.　Some Types. (According to Horn, W. and Kahle, I. (1935-7). Über entomologische Sammlungen, Entomologen & Entomo-Museologie. *Ent. Beih. Berl.-Dahlem* **2-4,** i-vi, 1-536.)

DYAR, H. G.　Historic material. (According to Horn, W. and Kahle, I. (1935-7). *Ent. Beih. Berl.-Dahlem* **2-4,** i-vi, 1-536.)

EARLY, W.　Crayfish females with ova in different states (1862).
　Specimens of female crayfish with eggs and young just hatched (1863).

EAST INDIA COMPANY　Collection of insects formerly in the Museum of the East India Company, presented by Curators of the Indian Institute, Oxford (1899). (The British

Museum (Natural History) had first choice including Types, the Indian Institute second selection.)

EASTOP, V. F.   119 microscope slides of British Aphididae (1949).

EBNER, R.   *see* HANITSCH, K. R.

EDDOWES, A.   Rare Papilio from the Upper Amazon (1889).
Apterous *Tettix* from the Upper Amazon (1891).

EDGAR, —   Coleoptera taken in neighbourhood of Oxford (1859).

EDWARDS, F. W.   *see* VERRALL, G. H. and COLLIN, J. E.

ELLIOTT, E. A.   Several rare and interesting forms of butterflies from British East Africa (1906).
Six examples of the Pierine butterfly *Euchloe charlonia* from Lanzarote, Canary Islands (1911).
Seven males of *Papilio dardanus* from Nairobi (1912).
Three males of *Danaida chrysippus* f. *alcippus* from Palma, Canary Isles (1913).

ELLISON, R. E.   11 butterflies from Abyssinia (1938).
118 butterflies from South America, mainly Panama (1949).
677 butterflies from Abyssinia (1951).

ELTON, C. S.   Insects collected on Oxford University Expeditions to Spitsbergen, Lapland, and Hudson Strait, 1921–31, also some from Norway, 1921.
Many small donations to British Collection (1940–66).

ELTRINGHAM, H.   An example of the mimetic butterfly *Crenidomimas concordia* from the Johnston Falls, north-east Rhodesia (1906).
Coleoptera and Rhynchota from Brazzaville, French Congo (1910).
126 Lepidoptera and 11 other insects from neighbourhood of Zurich and Lakes Lugano and Maggiore (1925).
His complete apparatus for low-power photomicrography, 200 lantern slides illustrating his researches, series of microscope slides, many of unique value, models of special structures of insects, his reagents, some rare and of considerable value; large number of offprints and a few books for the Library, also 52 butterflies and one dragonfly from Kumasi, Gold Coast. (Presented by Miss Eltringham, 1942.)

ELWES, H. J.   Five species of Satyrid butterflies from Sikkim, India (1887).
Pair of *Colias hecla* from Greenland and 99 specimens of the genus *Erebia* from Siberia, east Carinthia, Italian Alps, Tyrol, Altai, France, Switzerland, etc. representing nearly all the species (1899).
Butterflies from Canada (1899, 1900).

ENOCK, F.   Many slides, mainly Hymenoptera, prepared by F. Enock, given to J. O. Westwood in return for identifications. He also donated a few *Stylops* (1876). Enock gave up a lucrative career as engineer draughtsman to become 'Preparer of Microscopic Objects'.

ENTOMOLOGICAL CLUB   Part of a very old and historic collection, *c.*1826, presented by members of the Club (1927). Contains Diptera (except Types which went to British Museum (Natural History)), Hymenoptera, and Types of Lepidoptera (moths). The specimens are without data, but are of historic interest because they were identified by some of the most distinguished authorities of the nineteenth century; mainly incorporated in general collection.

ENTOMOLOGICAL SOCIETY OF LONDON   Insects of all orders purchased at sale of collections *via* J. O. Westwood (1857); others at sale, lot nos. 39, 41, 93, 110, 125–7, £2 (July 1863).

A box with nest of *Eucheira socialis* and nest of other insects, presented by Westwood (1863).

ESENBECK, C. G. NEES VON   68 syntypes of Chalcidoidea and 18 of Proctotrupoidea, described by him in his *Hymenopterorum Ichneumonibus affinium monographiae*, 1834, also six Cynipoidea. According to Westwood's diary he 'Received Nees von Esenbeck's Chalcid.ᵃᵉ from Bonn' in September 1836; collection otherwise destroyed in two world wars. The extant syntypes were discovered and have been arranged by M. W. R. de V. Graham, who intends to publish a list, with notes on the collection.

ESSEX, LADY   Cabinet *via* Hope Deed of Gift (1849).

EVANS, F. C.   *see* VEVERS, H. G. and EVANS, F. C.

EVANS, H. SILVESTER   Series of butterflies from Vanua Balavu, East Fiji (*c*.1924).
    Insects from the Congo-Zambesi watershed, including a group illustrating mimicry, collected 1927-9, also tubes of ant-spider mimics and various termites and associata (n.d.).

EVANS, J. W.   18 primitive species of Isoptera, Thysanoptera, Plecoptera, Hemiptera, and Mecoptera from Tasmania (1946); Isoptera and Thysanoptera in spirit, rest pinned.

EVANS, R.   Insects, chiefly butterflies, from British Guiana (1903).
    2157 butterflies collected by him during the Skeat Expedition to the Siamese Malay States, 1899 (1948).

EVETTS, A. F.   Collection of Lepidoptera, containing new species, from Balehonnur, Western Ghats, Mysore Province (1915-22).

FABRICIUS, J. C.   Specimens named by Fabricius in the cabinets of J. Francillon and J. Lee *via* F. W. Hope.
    A number of Iconotypes figured in W. Jones' *Icones* (see Appendix A).
    Type of *Aphelocheirus aestivalis* Fab.—original description states that it was in the Bosc Collection.

FAIRMAIRE, L.   Various insects by exchange with J. O. Westwood (1849).

FARGEAU, COMTE DE ST   *see* LEPELETIER, A. L. M., COMTE DE ST FARGEAU.

FARMBOROUGH, R. W.   Series of Lepidoptera from Trinidad (1915-22).

FARN, A. B.   Lepidoptera from various British localities (1914).

FARQUHARSON, C. O.   Insects from Southern Nigeria, 1914-18 (1915-22); donor lost in sinking of *Burutu*, 1918. (His observations and collections are fully described in *Trans. ent. Soc. Lond.* **1921,** 319-531.)

FAUST, J.   *see* BADEN, F. and SOMMER, M. C.

FAUVEL, A.   Specimens of *Cossyphodes bewickii* from Cape of Good Hope and *Actocharis marina* from France by exchange (1885).

FEATHER, W.   Collection of Somaliland moths, containing many new species (1915-22).
    Butterflies from Moa and Kibwezi, Tanganyika Territory (1923).

FELDER, —   [Vienna]   Insects of various orders *via* J. O. Westwood (1863).

FERGUSON, W.   Larvae and pupae of the 'cocoa nut' moth, with leaflets of the tree destroyed by the former, showing their mode of feeding; also *Coccus*, thrips, and spiders found on the 'cocoa nut' tree (1867).

FFENNELL, D. W. H.   Collection of over 16 000 British Lepidoptera, 500 Trichoptera,

75 Neuroptera; 356 microscope slides, nearly all genitalia preparations of Lepidoptera, and album of leaf-mines. (Presented by Mrs Ffennell, 1978.)

FILLEUL, P. R.   Lepidoptera from W. Maji Forest Reserve, Kericho, Kenya (1930).

FITCH, A.   About 800 species of North American insects (Coleoptera and Hemiptera) named by Fitch, *via* J. O. Westwood (1857).

FLETCHER, T. B.   54 mounted specimens of named termites from India and Ceylon (1912).

FLOWER, —   Crustacea (*c.*1869).

FLOWER, SIR W.   Five specimens of Homoptera, one *Curculio*, and one cocoon from Madagascar purchased by E. Ray Lankester (1894, £4).

FONSCOLOMBE, BARON E. L. J. H. BOYER DE   311 Hymenoptera Chalcidoidea collected by him around Aix-en-Provence, including some syntypes described in his *Monographia Chalciditum, Galloprovinciae circa Aquas Sextias degentum*, 1832; 24 Proctotrupoidea, 11 Cynipoidea (some syntypes), and 16 Bethyloidea including some Types described by Westwood; also some other insects. These were sent to Westwood in December 1836. A few other syntypes of the above groups exist in the Paris and Geneva museums, but the rest of his collection is destroyed. The collection has been arranged by M. W. R. de V. Graham, who is preparing a catalogue for publication.

FONSECA, E. C. M. D'ASSIS   166 predators with prey, mainly from Devon and Somerset (1948).
   132 Diptera and prey from Gloucester, Somerset, and Devon (1950).
   124 Diptera and prey including two species of flies recently added to British List (1951).

FORBES, C.   Shells and several crabs from Vancouver Island (1866).

FORBES, H. O.   Small series of insects from Peru (1919).

FORD, E. B.   Moths from Thursby, Cumberland (1923).
   11 butterflies and six Zygaenid moths from Isle of Canna, and seven butterflies from Malta (1947).
   See also DOWDESWELL, W. H. and FORD, E. B. (1939).

FORD, REVD H. D. and E. B.   Cumberland Lepidoptera (1915–22).

FORD, L. T.   106 specimens, comprising 30 named forms, of the moth *Peronea cristana* and 63 other Micro-Lepidoptera from various British localities (1953).
   153 specimens of forms of *Peronea hastiana* from Glamorgan and six *Nephopteryx* from Sussex (1955).
   British Micro-Lepidoptera new to or poorly represented in the Collections (1959, 1960).

FOREL, A.   *see* ROTHNEY, G. A. J.

FORREZ, M. A.   Beetles from Vancouver Island and California (July 1880, 6*s.* 6*d.*).

FÖRSTER, A.   230 species of German Hymenoptera named by him and given to F. W. Hope, mainly Chalcidoidea and Proctotrupoidea, some very probably syntypes; also some Formicoidea, Ichneumonoidea, and Cynipoidea. (Presented by F. W. Hope, August 1857.)

FORSTER, E. F.   Trapdoor spiders from Ceylon (1883).

FORTNUM, C. D. E.   Various Australian insects sent to Hope (1840–5).

FORTUNE, R.   *Coccus pela* the wax coccid of China from the Province of Chekiang (September 1857).

Exotic insects collected in China (purchased from S. Stevens by F. W. Hope, June 1858).

A Papilio collected in Japan (S. Stevens, July 1863, 6s.).

FOSTER, LIEUT. C. A.   Butterflies chiefly from neighbourhood of Freetown, Sierra Leone (1912).

Collection of insects, mainly butterflies, from Sierra Leone (presented by his father, 1914).

FOSTER, W.   Lepidoptera collected at Sapucay, Paraguay, 1903–5 (presented by Dr C. Hose, 1928).

FOUNTAINE, MISS M. E.   Six Syntomid moths, bred in the Durban district (presented by H. Rowland-Brown 1914).

Series of *Danais chrysippus* from Tenerife, Nigeria, and Cameroons, collected 1925–6 (n.d.).

11 butterflies from Budonga Forest, Uganda and seven from Madagascar (1934).

FOX, MISS M.   Insects of various groups from Pa-Ta-Ch'u, Western Hills, 12 miles (20 km) west of Peking (1926).

FRAMPTON, REVD R. E. E.   23 butterflies from the Chateaubelair district, St Vincent, West Indies, 1895 (1926).

FRANCILLON, J.   Cabinet *via* Hope Deed of Gift (1849). An undetermined Prionid beetle in Longicorn cabinet 2, dr. 36, bears a label in Westwood's handwriting 'This letter F. is in Kirby's handwriting & stands for Francillon at the sale of whose collection it was bought'; this is the first specimen to be definitely identified as out of the Francillon Collection, although there should be others.

The collection contained Haworth Types of Lepidoptera (see E. R. Bankes letter 13.6.1895 in Haworth file).

'Francillon's Catalogue' of 'Labels for Cabinet of the names of insects described by Fabricius' in Hope Library.

FRASER, LIEUT.-COL. F. C.   223 butterflies from South India including the very rare Satyrine *Parantirrhaea marshalli* (1937, 1941, 1942).

878 specimens of British Neuroptera and allied orders (1956).

281 specimens of British Trichoptera and Type and Allotype of the dragonfly *Cordulia linaenea* Fraser 1956, nov.nom. for *C. aenea* L. 1746 (1958).

FRASER, CAPT. J.   *see* DONCASTER, L. (1927).

FREEMAN, R. B.   Insects representing most of the larger Orders from Barra in the Hebrides, also specimens of *Bombus* and *Psithyrus* from Irish Free State and Hebrides (1935, 1937).

940 insects, mainly bumblebees, from Scotland and South Wales (1940).

159 insects from Co. Cork, Southern Ireland (1946).

32 insects from Co. Cork and Alderney, Channel Islands (1948).

FRENCH, G. H.   Lepidoptera from Illinois sent to J. O. Westwood by Miss Ormerod (November 1885).

FRIESE, H.   *see* JENSEN-HAARUP, A. C. *and also* MORICE, REVD F. D.

FRIVALDSZKY, E.   Coleoptera to Hope (1837–45).

FROGGATT, W. W.   Blattidae from various localities in Australia (1907).

FRUHSTORFER, H.   *see* DONCASTER, L. (1927) *and also* BURR, M. (1897).

FRY, T. B.   Indian Asilidae from neighbourhood of Poona (presented by J. W. Yerbury, 1902).

FRYER, J. C. F.   Lepidoptera from Aldabra (*c*.1912).

Whole of the material employed in his work *Investigation by pedigree breeding into the polymorphism of Papilio polytes* Linn. (*Phil. Trans.* (B), 1913, **204,** 227–54) (1915–22).

230 butterflies from Aldabra, 14 from Seychelles, and two from Assumption Island (1947).

Fyson, Mrs D. R.   Series of *Papilio polytes*, males and two mimetic forms of female, taken with their models near Madras City (1925).

Galton, J. C.   Insects from Atchin, Sumatra (1875).

Gammie, J. A.   Stingless species of honeybee with bread and comb, from Darjeeling, India (1887).

Garde, P. de la   Rhynchota Hemiptera and Homoptera from South Africa, Raro-tonga, Fiji, Sydney, Hobart, Queensland, Brazil, and Montevideo, and Lepidoptera from various European localities (1909).

20 Lepidoptera from Cooktown, northern Queensland and four butterflies from Thursday Island (1910).

Garrett-Jones, C.   About 1700 butterflies from the Alps and Mediterranean Europe (1974).

Gaye, J. A. de   Collection of insects, chiefly Lepidoptera, from Lagos (1912, 1913).

Geale, —   Various insects from Mexico and five bottles of Crustacea in spirits from Mr Cuming's collection (1866, 17s. 9d.).

Gebler, F.   Coleoptera from Siberia to Hope (1828–31).

Geldart, W. M.   Lepidoptera from Greece and Switzerland (1901–3).

Collection of British Lepidoptera (presented by Mrs Geldart, 1922 together with many books).

Gene, C. G. [Gené, J.]   Insects from Sardinia to Hope (1837).

Germain, P.   For Coleoptera collected in Chili, *see* Depuiset, A. (1866).

Germar, E. F.   Insects to Hope (1829–31).

Gerrard, E.   A few insects (purchased from Higgins, 1866).

16 small insects from Abyssinia, Hymenoptera, Hemiptera, etc. (January 1878, 10s.).

Insects from Abyssinia and the Argentine (April 1878, 13s. 6d.).

Butterflies (April 1889, £2. 4s. 2d.).

Cetoniidae and other Coleoptera from the Congo and Libellulidae etc. from Ceylon (1891, £4).

Various unspecified insects were also purchased between 1881 and 1892.

Giard, A.   *Typhlocyba hippocastani* and its two parasites (1889).

Gibbs, A. E.   Butterflies from British Honduras (1915–22).

Gilson, G.   Insects, chiefly butterflies, from Fiji Islands (1899).

Giraud, J. E.   Specimens of Hymenoptera: *Xyela, Paxyllomma, Cynips rosae, Tiphia, Nitela, Ampulex* (presented by J. O. Westwood, 1863).

Gladstone, H. S.   Series of Lepidoptera from Ibadan, near Lagos; butterflies and dragonflies from Calabar (1906). Many injured by pests, but some repaired and added to collections.

Godman, F. D.   12 Central American Phasmidae, including six syntypes described by Brunner von Wattenwyl (1908).

6625 insects (Coleoptera, Rhynchota, and Hymenoptera) from Central America, duplicates from material used for *Biologia Centrali Americana* (1909).

GODMAN, F. D. and SALVIN, O.   Over 16 000 butterflies chiefly from Central and South America, but also including a large number from other parts of the world (1896). Contains specimens of historic interest collected by H. W. Bates (Brazil), T. Belt (Nicaragua), A. R. Wallace (Malay Archipelago), and more recently G. C. Champion, H. H. Smith (Central America), and C. M. Woodford (Solomon Islands).
  This is the 'second set' of the great Godman–Salvin Collection and is incorporated in the general collection.

GOFFE, E. W.   Collection of British Diptera of the families Tabanidae and Syrphidae, also microscope slides, reprints, and papers dealing with Syrphidae (1952).

GOODRICH, E. S.   Insects, chiefly butterflies, from Ceylon (1899).
  Insects of many orders from Madeira (1901).

GORDON, C. J. M.   Lepidoptera and other insects collected in southern Nigeria (1901, 1902, 1903).

GORHAM, REVD H. S.   Purchased at auction, lot 286, 76 specimens (52 species) of Cleridae (March 1877, 3s. 6d.).
  Four examples of Erotylidae from the Congo and New South Wales, named by donor (1903).

GOSS, H.   327 specimens of British Lepidoptera, captured and bred (1898).

GOUREAU, C. G.   Five Types of Chalcidoidea described by him in his *Mémoire pour servir à l'histoire des Diptères*, 1851.
  A few other syntypes may exist in Paris, otherwise his material appears to be lost.

GRAHAM, M. W. R. DE V.   104 Types of Hymenoptera Chalcidoidea described by donor (1981).

GRANT, W. R. OGILVIE-   *see* OGILVIE-GRANT, W. R.

GRAVENHORST, J. L. C.   Some Types of Hymenoptera Ichneumonidae described in his work *Ichneumonologia Europaea*, 1829, from material supplied by F. W. Hope. Also insects by exchange with Hope (c.1828–30).

GREEN, E. E.   Microscope slides of insects (1891, £3).
  Six cocoons of Tineid moth *Epicephala chalybacma* from Peradeniya, Ceylon (1912).
  Microscope slides; also phials of Coccidae, some of which are paratypes (1941). Includes some Coccidae sent by W. J. Hall from Egypt in 1922.

GREEN, J. S.   Specimens illustrating the transformation of a case-making *Aphrophora* in Ceylon (1885).
  Aphis artocarpi, a new species which infests the breadfruit tree in Ceylon (1889).

GRENSTED, REVD L. W.   Many specimens of British insects, notably Trichoptera, Neuroptera, and Diptera (1932–57).
  43 Swiss Diptera (1934).
  Two boxes of old microscope slides (presented by Mrs Grensted, 1964).

GRIFFITHS, G. C.   Insects from world-wide localities (1894–1906).
  Lepidoptera from north Shetland and Scotland (1899).

GROSE SMITH, H.   *see* SMITH, H. GROSE.

GUÉNÉE, A.   Selection of his South American moths including Types in Noctuid collection.

GUÉRIN-MÉNEVILLE, F. E.   76 insects to Hope (1841).
  Box of specimens of *Bombyx*, *Saturnia*, and *Cynthia* with cocoons and silk spun and manufactured (May 1861).

Small collection of Crustacea in spirits, chiefly exotic, from the Voyages of Souleyet, Gaudichaud, etc., contained in 51 bottles, comprising many new genera and the unique Type of *Cystisoma neptunus*, also 27 bottles containing illustrations of transformations of insects, chiefly larvae of North American Coleoptera (May 1863, £3, carriage 5*s*.).

GUILDING, REVD L.   Two cabinets *via* Hope Deed of Gift (1849).

GUNTHER, R. W. T.   Insects captured in the neighbourhood of Urumiah, north-west Persia (1898).
174 Lepidoptera from Cameroons from cabinet purchased from Stevens (1936).

GURR, J.   30 rare or less common British spiders, through G. H. Locket (1976).

HAAN, W. DE   *see* RIJKSMUSEUM VAN NATUURLIJKE HISTORIE, LEIDEN.

HAARUP, A. C. JENSEN-   *see* JENSEN-HAARUP, A. C.

HAINES, F. H.   Collection of spiders in 140 jars (presented by Mrs Haines, 1948).

HALIDAY, A. H.   Two specimens of *Nebria borealis* from Lough Neagh and three of *Pseudopsis sulcata* from Holywood to Hope (September 1846).
Part of his collection of Chalcidoidea Mymaridae through exchange with J. O. Westwood (1886) (see Graham, M. W. R. de V., 1982, *Proc. R. Irish.Acad. (B)* **82 B** (12), 191–2, 237–9).
The Collections contain a few syntypes of other Chalcidoidea collected by Haliday and described by F. Walker, and a few described by Haliday himself; also Irish Coleopterous larvae sent to Westwood in 1856. *See also* DALE, J. C. and C. W.

HALL, W. J.   *see* GREEN, E. E. (1941).

HAMM, A. H.   Curious form of *Heliconius erato* (1896).
British Diptera and two boxes of Agromyzid leaf-mines, also a vast collection of insects and their prey, fossorial wasps, parasitic Hymenoptera bred from known hosts, and other bionomic material over a number of years (1897–1942). The collection is rich in Nematocera, minute Empididae, bred Sarcophagidae, Tephritidae, and Agromyzidae. Bionomic collections include Strepsiptera and their hosts, bred parasites and insects bred from moles' and birds' nests; also Aculeate Hymenoptera.

HAMM, TPR. E. E.   Insects collected in the Transvaal (through A. H. Hamm, 1904).

HAMMOND, H. E.   433 blown larvae of British Lepidoptera, mainly moths, representing over 200 species (purchased 1960–2, 1964).

HAMMOND, H. E. and SMITH, K. G. V.   263 specimens of parasitic Hymenoptera, including the first bred specimens of *Enicospilus repentinus* (Holmgren) and a hitherto undescribed species of *Eulophus* (1956).
50 parasitic Hymenoptera bred from known Lepidopterous hosts (1957).

HAMPSON, SIR G. F.   280 butterflies from the neighbourhood of Courmayeur, Italy (1905).

HANBURY, D.   Specimen of wild silk from San Salvador (1863).

HANCOCK, G. L. R.   Lepidoptera from Uganda, including the rare *Mylothris ruandana* and a series of *Acraea rahira* not previously recorded from Uganda (1933).

HANCOCK, J. L.   28 Tetriginae and seven Blattidae from Mexico, Illinois, Indiana, Massachusetts, Texas, and Wisconsin, all determined by donor (1907).

HANDSCHIN, E.   *see* HANITSCH, K. R.

HANITSCH, K. R.   Insects from Selangor, Federated Malay States and series of *Papilio polytes* from Singapore Island and the mainland (1915–22).

Many specimens of Orthoptera, especially Blattidae, from duplicates of collections sent for determination whilst he was working in the Hope Department, by:

L. Chopard, Java and Borneo. R. Ebner, East Sumatra. E. Handschin, Java and northern Australia. Sven Hedin (*via* Sjöstedt), China. J. Hewitt, South Africa. E. Jacobson, west Sumatra. C. B. Kloss and H. H. Karny, south Annam, Mentawi Island, Sumatra. W. R. Ladell, Siam. E. Mjöberg, Deli, north-east Sumatra. E. Modigliani (through Prof. Gestro) and O. Beccari, Sumatra, Mentawi, and Engano. H. M. Pendlebury, Malay Peninsula, British North Borneo. H. C. Siebers, south Sumatra. Y. Sjöstedt, China and Amazon. N. Smedley, west Sumatra. H. J. Snell and H. P. Thomasset, Rodriguez.

Fine set of offprints containing all the more important papers on Orthoptera over past 50 years (1938).

HANSARD, — Nests of *Mygale fodicus* from Corfu, and nests of trapdoor spiders (1862).

HARDWICKE, MAJ.-GEN. T. Some insects from Nepal to F. W. Hope (*c.*1833).

HARDY, G. H. Series of insects from Katoomba, New South Wales (1954).

HARRIS, T. W. Box of North American insects to J. O. Westwood (1836).

HARRISON, A. H. Butterflies from the neighbourhood of Lake Victoria Nyanza (1903).

HARRISON, G. HESLOP- *see* HESLOP-HARRISON, G.

HARRISON, J. W. HESLOP- *see* HESLOP-HARRISON, J. W.

HARVEY, REVD J. H. Collection of spiders from British localities (1937).

HARWOOD, P. 127 specimens of Hymenoptera Aculeata, Tenthredinidae, and Diptera, mainly from Scotland (1955).

533 specimens of Staphylinid beetles of the genus *Atheta*, 747 sawflies, and a few other insects of special interest from various British localities (July 1956).

958 specimens of British Hemiptera and 220 of British Ichneumonidae and 48 Braconidae (December 1956).

Collection of British Coleoptera and Diptera, consisting of six cabinets containing 85 drawers of Coleoptera (approximately 100 000 specimens) and 20 drawers of Diptera (approximately 13 000), also 12 store-boxes of Diptera and 10 of Coleoptera (presented by Mrs Harwood, 1957). [The Hemiptera and Hymenoptera are in the British Museum (Natural History).]

Mrs Harwood also presented a microscope for use by students in the department.

HAUSBURG, C. B. *see* MACKINDER, H. J. (1899).

HAVILAND, MISS M. D. Specimens including some paratypes selected from a collection of Membracidae made in British Guiana in 1922 (presented by Newnham College, Cambridge, 1924).

HAWORTH, A. H. Cabinet *via* Hope Deed of Gift (1849) and cabinet filled with exotic insects of various orders *via* J. O. Westwood (1857). The collection contains British Lepidoptera, including some types with triangular labels; a few parasitic Hymenoptera, bearing tiny blue square tickets indicating British origin. The Departmental Report of 1910 also mentions British Staphylinidae bearing Haworth labels.

The Haworth collection was catalogued by Westwood, and sold in 1834; the sale realized £552. 12*s*.

See also FRANCILLON, J.

HAYES, A. J. Insects from Lake Tana, Abyssinia (1903).

HEALE, W. H. 146 butterflies from Zululand, Natal, and Cape Colony (1903).

HEARSEY, MAJ.-GEN. SIR J. B.   Cabinets containing Indian Lepidoptera, Coleoptera, Hymenoptera, Hemiptera, Orthoptera *via* J. O. Westwood (1857). [Approximately 1450 Lepidoptera and 300 Coleoptera captured in India, stated to be Professor Westwood's private collection, were purchased from his niece, Miss Swann, in 1895.]

HEATH, R. H.   A few butterflies collected at Colon, Panama, and in Trinidad (1927).
Coleoptera collected by the donor between 1886 and 1891 when at school at Ackworth, West Riding, Yorkshire (1927).
363 British Lepidoptera (presented by A. R. Heath, 1927).

HEAVISIDE, J.   Cabinet *via* Hope Deed of Gift (1849).

HEDIN, S.   *see* HANITSCH, K. R.

HERNANDEZ, M.   Series of Coleoptera and other insects from neighbourhood of Mahon, Menorca (1900).

HERRICH-SCHÄFFER, G. A. W.   Thirty specimens of Proctotrupoidea of which six probably represent Types, sent to Westwood (1842).

HESLOP-HARRISON, G.   451 aquatic Coleoptera from Outer Hebrides (presented by J. W. Heslop-Harrson, 1940).

HESLOP-HARRISON, J. W.   Insects from the Hebrides including 1050 aquatic Hemiptera in spirit (1939–50).

HEWITSON, W. C.   33 exotic Lepidoptera, chiefly butterflies (1859).
12 specimens of *Hipparchia* (1863).
Butterflies by exchange (1864).
Small collection of butterflies from Old Calabar, Waigiou, and Venezuela (1865).
Argynnis Diana Cramer, West Virginia '(rariss.ᵃ)' (1867).
19 *Catagramma* from his duplicates (1871, £1).
Lepidoptera, including some from Rogers' Gaboon Collection, and Coleoptera (£4. 9s., and 17s. 6d., 1873).
Specimens from Rogers and Chesterton collections, Wallace Malay collection and Hewitson duplicates (1874, £13. 5s. 6d.).
Papilio zalmoxis (1876, 10s.).
Insects from Lake Nyasa, Mozambique (1877, 13s.).
Many butterflies purchased between 1871 and 1875.

HEWITT, J.   25 Blattidae and three Mantidae from Sarawak (1905).
Some South African Orthoptera through K. R. Hanitsch and a few Coleoptera (1906).

HICKS, J. B.   Lepidoptera of the world (presented by A. S. Hicks, 1934).

HICKSON, S. J.   Coleoptera from Celebes (1885).
Various Lepidoptera, etc., from Celebes (1887, 1888).

HIGGINS, MR (Late of S. STEVENS)   Insects of all orders (purchased 1866–76, £9. 2s., £7. 15s., £8. 10s., 19s. 6d., £5. 13s., £5. 4s., £5. 18s. 10d., £4. 16s., £9. 19s. 6d., £8. 19s. 9d., etc.).

HILL, G.   Two jars of spiders from Christmas Island, Indian Ocean, [1937]–9 (1941).

HINDE, S. L. and MRS HINDE   Butterflies from British East Africa (1900).
Insects of various orders, especially Lepidoptera, from British East Africa (1901–4, 1906).
Contributed many other specimens and interesting notes from Kenya Colony.

HOCKIN, J. W.   Large series of butterflies and other insects from Travancore (1915–22).

HODSON, SIR A.   Large collection of butterflies from Abyssinia, also some from the
   Sudan (1925–7).
      Butterflies and moths from Sierra Leone (1934, 1935).

HOFER, C.   *see* DONCASTER, L. (1927).

HOLLAND, W.   Insects of all orders from various localities both British and foreign
   including some Lepidoptera collected by Col. C. Swinhoe (1894–1903).
      Collection of British Carabid beetles (1907, 1910).

HOLMES, MISS M. G.   Lepidoptera, Hymenoptera, and Coleoptera from Virden, Manitoba (1899).
      Set of 80 insects from near Laurier, Manitoba (1901).

HOOD, CAPT. THE HON. G.   180 butterflies from Ashanti (1899).

HOOKER, SIR J. D.   Specimens of 'cocoa nut' fruit from Cuba, infested by gigantic
   species of *Bruchus*, four beetles from Caracas, cocoons and chrysalids of the Chinese
   fishing-gut moth (1867).

HOOKER, SIR W. J.   '*Bruchus* (*Caryoborus*) *Bactris?* Larva fr. Palm nut Bahia Sir W.
   Hooker 18.8.53' (J. O. Westwood, n.d.).

HOPE, REVD F. W.   Schedule appended to Deed of Gift, 4 August 1849 (with minor
   amendments to original spelling):

Cabinets	Double Drawers	Method of arrangement	
1	Dufours Cabinet 30 double drawers	Orthoptera	60 drawers
2	Locust Cabinet of 9 ditto ditto	ditto	18
3	Humphries Cabinet 18 ditto ditto [?Humphrey]	Curculio	36
			114

	Single drawers		
4	Hope's Magazine Cabinet	Coleoptera all orders	62
5	Hope's New Magazine Cabinet		58
6	Drury, Dru: Cabinet		60
7	Haworth's Cabinet	Lepidoptera	58
8	Lamellicorn Cabinet (Dufours)		30
9	Count Billberg's Cabinet		24
10	Lee's Cabinet	Lamellicorns	27
11	Cetonia Cabinet and Buprestes		24
12	Longicorn Cabinet		24
13	Dr. Heaviside's Cabinet 26 corkes and 4 large		32
14	English Coleoptera Cabinet		42
15	Sir Patrick Walker's Cabinet		32
16	English Hemiptera and Hymenoptera		16
17	Latham's Cabinet	Neuroptera	22
18	Lady Essex' Cabinet	Orthoptera	20
19	Sowerby's Cabinet	Carabidae	44
20	Orthoptera Cabinet		13
21	Raddon's Cabinet (English)		30
22	Foreign Hymenoptera Cabinet		8
23 and 24	Lansdown Guilding's Cabinet 14 and 14		28
25 and 26	Rose Wood Cabinets 12 and 12		24

27	Heteromera Cabinet	24 drawers
28		
29	Three Indian Cabinets 10 drawers each	30
and		
30		
31	Longicorn Cabinet in table	20
	Table Longicorn Cabinet Continued	16
	Francillon's Cabinet	36
	Dillenius Cabinet Fossils Shells Minerals &c	13
	Cimex Cabinet (Scuttellerida)	9

The Schedule also included boxes of fossil insects, ambers, resins, fossil Crustacea, Crustacea and molluscs in spirit, shells, about 200 solanders containing plates and original drawings relating to natural history, 49 solanders containing portraits of naturalists, 18 portfolios of topography, and the library which consisted of about 1800 volumes.

Collectors mentioned in the Schedule are also listed under their respective names. Donations continued to arrive after 1850, of which there is no proper record, but from 1857, when Westwood started his journal, he listed the following gifts:

1857	Coleoptera from Transylvania, Austria, etc.
	Insects from Nice.
	British Crustacea; gigantic specimens of crabs and lobsters and other Crustacea, dried and also in spirit, from the Mediterranean.
	Fossil insects from Aix-en-Provence.
1859	Coleoptera from New South Wales, Chili (or Peru), and Brazil.
1860	Eggs and bird skins.
1861	Plaster casts of birds' and reptiles' footsteps in new red sandstone from Cheshire.
	Crabs, fossils, and fossil shells from Suffolk.
	Cast of head of a Saurian.
	Ammonite and fossil wood from Van Diemen's Land.
	Goosander, ducks, etc.
	Crocodile from the Nile.
	Eggs of Mute Swan, Canada Goose, Guillemot, etc.
	Skulls of the Albatross, Heron, Pelican, Toucan, and Condor.
	Four varieties of Tortoise.
	Skull of Porpoise, head of Turtle.
	36 bottles containing fish, snakes, reptiles, etc.
1862	Lynx, birds, fishes, tortoises, sea hedgehogs, and a snake in spirit.
1862	'Contents of Boxes &c.' ...

    No^d   1, 2, 3. Portraits & Books.

          4. Casts, part of Mummy Case, Portraits Glazed & Framed.

          5. Topography &c.

          6. Insects &c.

          7. Insects in Bottles &c.

          8. Minerals.

          9. Minerals & Fossils.

          10. Fossils.

          11. Fossils.

          12. 1 Cast.

          2 Portfolios—Portraits & Topography

          2 small Cabinets of Insects.

1 Box with Trays for Portraits.
1 Store Insect Box with Portraits.
1 Small Portfolio   do.
1 Solander with paper for mounting Topography.
Total No. of Portraits about 17,900'

HOPE-SIMPSON, J. F.   225 butterflies from south-west Sudan (1940); 32 butterflies from Northern Rhodesia (1950).

HOPKINS, G. H. E.   Butterflies from Samoa and Tonga (*c.*1925).

HOPKINSON, E. M.   Insects from the neighbourhood of Bathurst, Gambia (1906).

HORN, —   Crustaceans collected in the Red Sea (sent by Mr Lord to J. O. Westwood, 1870).

HORN, G. H.   *Platypsylla castoris*, parasite upon American beaver (1888).
    See also ANDREWES, H. E. and F. W. (1900).

HORNE, A.   British butterflies (1923, £3. 6*s.* 9*d.*).

HORNIMAN, F. J.   Some insects to Westwood (1879).

HORNSEY, J. F.   40 insects of various groups, chiefly Lepidoptera, from British North Borneo (1912).

HORSFIELD, DR T.   Lepidoptera from Java and Natal *via* J. O. Westwood (1857).

HOSE, DR C.   Coleoptera from Celebes collected 1895–6, including mounts with two or more specimens (1914).
    See also FOSTER, W. (1928).

HOTTOT, R.   Insects from French Equatorial Africa (1932).

HOWARD, L. O.   Moth and a beetle mimicking each other, captured in the Chiricahua Mountains, Arizona (1897).

HOWARTH, O. H.   Set of 80 Lepidoptera and one Coleopteron from neighbourhood of Oaxaca City, Mexico (1897).
    Insects of many orders, chiefly Lepidoptera, from Mexico and four Pierines from Tembabichi Bay, Lower California (1898, 1899).

HOWDEN, MRS A. T.   17 paratypes of species of the Curculionid *Pandeleteius* from Colombia, Venezuela, and Arizona, described by donor (1976).

HOWDEN, H. F.   14 paratypes of North American Scarabaeidae by exchange (1963).

HOWITT, G.   Insects, mainly Coleoptera, from Melbourne and Launceston, Van Diemen's Land to F. W. Hope (1842).
    ♂ and ♀ of *Dorcas howitta* Westwood from the mountains of Victoria, New Holland, and ♂ of *Rhyssonotus jugulans* Westwood (presented by J. O. Westwood, 1865).

HUBBARD, C.   Collection of stuffed birds of the Orkneys (purchased and presented by F. W. Hope, 1861).

HUDSON, G. V.   New Zealand Lepidoptera (1915–22)—damaged in transit.
    New Zealand beetles (1939–46).

HUGHES, MR   12 butterflies from Levuku, Fiji Islands (1878, 14*s.*).

HUGHES, REVD G.   Rare British moths (1917, £7. 10*s.*).

HUMPHRIES, — [HUMPHREY, G.?]   Cabinet *via* Hope Deed of Gift (1849).

HUNT, REVD A. P.   389 Lepidoptera, almost exclusively butterflies, from various localities in North America (1904).

INDIAN INSTITUTE, OXFORD   *see* EAST INDIA COMPANY.

IMAGE, S.   British Lepidoptera (presented by Mrs Image, 1930).
Collection contains very rare or extinct butterflies and Curtis Types of the vars. *stonanus* and *ramulanus* of *Sarrothripus revayana*; collection incorporated, but one special cabinet has been retained to comply with her wishes; many books presented including some by Donovan, Stephens, Curtis, and Roesel.

IMRAY, MRS R.   Insects of various orders, mainly Coleoptera, from Madras, Travancore State (presented by Mrs Imray and Revd A. Thornley, 1903).

INGALL, T.   Selected Coleoptera and Lepidoptera (purchased from St Bartholomew's Hospital, 1927).

INSTITUT ROYAL DES SCIENCES NATURELLES DE BELGIQUE, BRUSSELS   Blattidae (1908).
Representative series of butterflies from Congo State selected from a collection exhibited to the King of the Belgians (1910).

JACKSON, A. RANDALL   see TAYLOR, E.

JACKSON, F. W.   Lepidoptera from Trinidad and other West Indian Islands, also miscellaneous Jamaican insects (1915–22).

JACKSON, W. HATCHETT   Caterpillars of British Nymphalidae in different stages of their transformations (1890).

JACKSON, T. H. E.   Vast collections of butterflies from Kenya, Uganda, Southern Rhodesia, Belgian Congo, and Abyssinia (1932–50).

JACOBSON, E.   Some Blattidae from Fort de Kock, western Sumatra, *via* K. R. Hanitsch (1926); insects, mainly Coleoptera, from Sumatra (1928).

JACOBY, M.   19 specimens of Lepidoptera from Tasmania (1897).
Two specimens of a species of *Nyctelia* (Coleoptera) from Patagonia (1898).
Coleoptera Phytophaga from various localities (1906).
See also ANDREWES, H. E. and F. W. (1900).

JANSON, O. E. (Natural History Agent)   Various insects purchased between 1863 and 1879, including some from north coast of New Holland (1866, £1); Nepal, Upper Amazon, New Holland, and Moluccas (1867, 14s. 6d.); one Lamellicorn, *Cyphelytra ochracea* Waterh., from Darjeeling (1875, 15s.).
Van der Poll collection of Carabidae and Orthoptera (purchased by Poulton, 1909, £40); *see also under* Chevrolat, L. A. A., *and* VAN DER POLL, J. R. H. NEERWORT.

JEFFERY, G. W.   162 butterflies from Mombasa (1948).

JEKEL, H.   Some Curculionidae *via* Revd W. Tylden (1875).

JENISON-WALWORTH, GRAF R. VON   Some insects to Hope (1834).

JENKINS, MAJ.   Lepidoptera from Assam *via* J. O. Westwood (1857); there are also some Hemiptera in the collection.

JENNER, J. H. A.   Six specimens of rare Coleoptera from Sussex (1895).

JENNESS, D.   Lepidoptera from Bwaidoga, Goodenough Island, New Guinea (1912).

JENSEN-HAARUP, A. C.   117 bees from Argentina; includes species determined by Friese and syntypes of 17 new forms described in *Die Apidae (Blumenwespen) von Argentina*, by H. Friese, 1908 (purchased 1908).

JERDON, T. C.   Indian insects to Hope (1846).

JERMYN, REVD E.   Specimens of *Galeodes* and *Cicada* from India (1890).

JERMYN, COL. T.   Indian butterflies and additions to mimicry series (1915–22).
Collection of North Indian butterflies (presented by Mrs Jermyn, 1928).

JOHNSON, C. F.   British moths including many rare species (1912, 1914, 1923).

JOICEY, J. J.   Butterflies from Africa, Oriental Region, and tropical America (1912), and central Madagascar, New Guinea, and the Andes north Peru (1913). *See also* PRATT, A. E., C. B., and F. B.
   Butterflies, some of them belonging to mimetic associations, from Africa, South America, and the Austro-Malayan Islands; the latter include a fine set of *Delias* from New Guinea and mimetic set from Tenimber Island, Timor Laut. (1915–22).
   Butterflies from S. Thomé Island, West Africa (1926).
   Residue of his collection freely offered to University (1934).

JONES, C. GARRETT-   *see* GARRETT-JONES, C.

JONES, W.   Coleoptera, Lepidoptera, and a few other insects, including two males and one female 'Large Copper' from his collection, probably the oldest existing specimens of this extinct butterfly; also four pieces of copal containing insects. (Presented by Dr F. D. Drewitt 1925 and 1929.) 44-drawer cabinet of British Lepidoptera (Dr F. D. Drewitt, 1931, £20). Kept separate.

JUNOD, H. A.   *see* DONCASTER, L. (1927).

KADEN, C. G.   Miscellaneous Lepidoptera mainly larvae and pupae (presented by J. O. Westwood, 1863).

KARNY, H. H.   *see* HANITSCH, K. R.

KAYE, W. J.   Lepidoptera, mainly butterflies, from central British Guiana (1902) and British East Africa, Peru, St Helena, India, British Guiana, Tibet, Switzerland, Italy, and France (1903, 1906, 1909, 1911).
   15 dragonflies collected on the canal banks, Georgetown, British Guiana (1911).
   South American butterflies (1915–22).

KEATINGE, G.   Collection of butterflies from Karwar, Bombay Presidency (1899).

KELLY, R.   Group of insects, taken on one day, visiting flowers of Eucalyptus at Healesville, Victoria (1915–22).

KENNY, J. S.   Six slides of *Trichobius caecus* (Diptera, Streblidae) from Tamana Cave, Trinidad (1970).

KENRICK, SIR G.   Pierine butterflies from New Guinea (1912, 1915–22).

KENT-LEMON, A. L.   Series of Lepidoptera from the Nuba Hills and south Sudan (1915–22).
   See also WILSON, CAPT. R. S.

KERSHAW, J. C.   Collection of butterflies from Macao and Hong Kong and other insects (1904–5).
   Coleoptera and a few other insects from Piroe, south-west coast of Ceram (1908).
   10 examples of the moth *Aulacodes simplicialis* and the Ichneumonid parasitic upon it from Lappa Island captured by donor and F. Muir, also 16 small moths taken at sea (1909).
   Small collection of butterflies from Queensland (1911).
   31 Asilid flies from How-lik Monastery, near Hong Kong (1913).

KETTLEWELL, H. B. D.   1711 specimens of British and foreign Lepidoptera, representative of his research interests (presented by Mrs Kettlewell, 1979); main collection in British Museum (Natural History).

KILLINGTON, F. J.   British Neuroptera and Trichoptera of special interest to the Oxford area (1932–3).
   22 Buprestidae, Passalidae, Scarabaeidae, etc., from Australia (1936).
   Collection of some 4500 insects presented in accordance with his known wish

(1957). Contains material used in the preparation of his Ray Society Monograph on the British Neuroptera, 1936-7. Mrs Killington also presented a series of microscope slides of dissections of Neuroptera which originally formed part of the main collection (1964).

Kirby, Revd W.   A portion of his collection (*c*.1858-60).
>Two of his specimens of *Eulophus damicornis* Kirby (Hymenoptera Chalcidoidea) in the Dale Collection.
>Type of *Bolbocerus proboscideus* purchased at sale of Entomological Society's British Collection for 1*s*. in July 1863.
>*See also* note under Latreille, P. A. (bees).

Kirkaldy, Miss J. W.   Beetles from New Guinea (1890).

Kloss, C. B.   Blattidae from Annam, determined by K. R. Hanitsch (1926).

Klug, J. C. F.   Some insects, including Pompilid wasps and bees and other Hymenoptera named by Klug.

Koch, L.   *see* Pickard-Cambridge, Revd O. (1917).

Kolenati, F. A.   A few Hemiptera in Hope-Westwood Collection.

Kollar, V.   Coleoptera and Hymenoptera sent to Hope (1845), some representing new species of Hymenoptera Chalcidoidea which were described by F. Walker in 1847.

Konow, F. W.   *see* Morice, Revd F. D.

Kowarz, F.   Diptera *via* Verrall-Collin Collection (1967).

Kraatz, G.   Four specimens of the Silphid beetle *Adelops Bonvouloiri* (1863).

Kricheldorf Collection   *see* Doncaster, L. (1927).

Kunze, G.   Insects, mainly Coleoptera, sent to Hope (1832 and 1836).

Kunze, R. E.   Insects of many orders from the United States, including over 2000 Coleoptera from Arizona (1910-13).
>Small collection of butterflies from Arizona (presented by his nephew Richard Kunze, 1929).

Lack, D.   *see* Cambridge University Expeditions.

Lack, D. and Venables, L. S. V.   Insects of all orders from Galapagos Expedition, 1938-9, including scorpions, myriapods, etc., in spirit; the 60 butterflies comprise all of the known species, with the exception of *Pyrameis* (1941).

Lacordaire, J. T.   'Mecomastyx montravelii—♀ Montrouzier, New Caledonia; Cnemacanthus?—(2) Buenos Ayres; n.g. Nyctelia aff. (2) Buenos Ayres' (1865).

Ladell, W. R.   *see* Hanitsch, K. R.

Lagace, C. F.   18 bred specimens (four slides and 12 pinned, paratypes) of *Mesidia nigra* (Hymenoptera: Aphelinidae) from California described by donor (1969).

Lamborn, W. A. S.   Large collection of Lepidoptera including bred material, and other insects and bionomic material from West Africa, ex-German East Africa, Tanganyika Territory, Nyasaland, Federated Malay States, China, Japan, Canada, Madagascar, and Comoro Islands (1910-42); books and offprints (1960).

Lampel, G. P.   Collection of Arachnids (presented by his parents, 1962). This material dates from about 1950 onwards and contains many specimens accompanied by precise locality and habitat data. It includes specimens from Britain, Europe (mainly France and Italy), Persia, Ethiopia, Canary Islands, and British

Guiana, and some collected jointly with J. A. L. Cooke in Turkey. Also books and offprints.

LANCE, I. H.   Insects from Surinam to Hope (1833).

LANCHESTER, W. F. and MINCHIN, E. A.   468 Lepidoptera, chiefly butterflies, from Banyuls-sur-Mer and Argelès-sur-Mer at the eastern extremity of the Pyrenees (1909).

LANG, REVD H. C.   More than 1500 European butterflies (presented by H. Rowland-Brown, 1914).

LANGHANS, J.   Hemiptera from Sydney and Brisbane, Australia, South Island, New Zealand, and Buitenzorg, Java, collected 1891–3, also some Coleoptera from Sydney (1896).

LANKESTER, C. H.   240 butterflies and 61 moths from Sudan, Uganda, and Rhodesia 1921 (1941); a few other insects from Cartago, Costa Rica.

LAPORTE, F. L. N. DE C. DE, COMTE DE CASTELNAU   998 exotic insects from Malacca, Lake N'Gami (South Africa) and British Caffraria (1862).

LATHAM, J.   Cabinet *via* Hope Deed of Gift (1849).

LATREILLE, P. A.   The following Latreille material came through J. O. Westwood (1857); the list has been prepared by M. W. R. de V. Graham:

1. Five drawers of dissections of insects, made by Latreille; in the first drawer is pinned the following note:

> Paris, le 26 Otbre 1832.
> Mon cher Monsieur,
> ne m'oubliez pas, je vous prie,
> auprès des entomologistes de votre pays
> et conservez-moi toujours une bonne
> part dans votre amitié.
> Latreille, prof au Mus.
> d'hist. natur.
> pour Monsieur Westwood.

2. Collection of bees (includes numerous Types). Two drawers in Hymenoptera cabinet in beetle-room; plus a number of specimens removed from above drawers by Dr Tkalců and now placed in a supplementary drawer of the Type Collection (Hymenoptera). At the top of the first drawer of Latreille bees is a note in Westwood's handwriting, reading:

'Latreille's collection of bees (purchased by Dejean and named in part by St. Fargeau) passed into the hands of Serville after whose death I obtained it in its present state I.O.W.'

Many specimens have labels in Latreille's handwriting. Some, however, have labels in the handwriting of Lepeletier (evidently those referred to above in Westwood's note 'named in part by St. Fargeau'). A few I have been able to determine as having come from Dejean's collection, and I have so labelled these. Against three specimens which bear tickets numbered 155, 173, and 146, Westwood has pencilled a note on the paper of the drawer 'Kirby Labels?'.

3. A few other small Hymenoptera (Chalcidoidea, etc.): 'From the Abbé Blondeau (being part of Latreille's original collection sold to Dejean and bought by the Abbé at the sale of the Collection of Dejean by Auction' (Westwood MS note).

LAWRENCE, SGT.   Insects, mainly butterflies, captured in Sierra Leone (purchased 1909).

Leadbeater, — Insects from India, mainly Coleoptera, purchased by W. Bainbridge for Hope (1841).

Le Conte, J. E. American Coleoptera to Hope (*c*.1835). Other insects (*c*.1842–6).

Lederer, J. Insects of various orders, many from Syria (presented by J. O. Westwood, 1863).

Lee, J. Cabinet *via* Hope Deed of Gift (1849): 'Insects with old labels with a thick black line are from Lee's Collection of Hammersmith often named by Fabricius himself' (Westwood MS note).

Leech, H. B. 176 specimens of Coleoptera from British Columbia (by exchange, 1932).

Leech, J. H. Lepidoptera from China, Japan, and Tibet (purchased from Stevens, 1901).
See also Doncaster, L. (1927).

Lefebvre, A. Insects to Hope by exchange (*c*.1831–5).

Leigh, G. F. Insects of various orders from localities in and near Durban, Natal (1904; 1906, £2. 2*s*.).
Bred series of *Papilio dardanus* and its forms *cenea* and *trophonius* (1904–10, 1924, some purchased, some presented).
Insects of various orders from Durban, Natal, etc. (1908, 1909).
Lepidoptera from Grand Comoro, Johanna, and Mayotta (1911, £2. 14*s*. 6*d*., £7. 15*s*.).
Lepidoptera from Natal (1923, £6; 1924, £2).

Leman, G. C. Miscellaneous Swiss insects (presented by H. St J. K. Donisthorpe, 1925).

Lemon, A. L. Kent- *see* Kent-Lemon, A. L.

Lepeletier, A. L. M., Comte de St Fargeau Many exotic bees, including some Types, *via* J. O. Westwood (1857).
See Latreille, P. A.

Le Quesne, W. 138 British Hemiptera Auchenorhyncha (1981).

Lester-Smith, W. C. Some moths from Trinidad (1925).

Leston, D. Over 5000 British Hemiptera–Heteroptera (1963).
Approximately 10 000 Hemiptera Pentatomidae of the world (1965).

Lethierry, L. F. Some Palaearctic Hemiptera. *See* Atkinson, E. F. T. (1895).

Lever, R. A. Butterflies, including many Euploeinae, from the Solomon Islands (1932–7) and Fiji (1946).

Lewis, G. W. Specimens of *Syntelia histeroides* (Lewis) from Japan and *Prostomis schlegeli* from Ceylon (1885).
Rare beetles of the families Rhysodidae, Carabidae, Cucujidae, etc., mainly from Japan (1887).
Specimens of *Paussus chevrolati* (1889).

Lewis, R. Box of insects from Van Diemen's Land to J. O. Westwood (1837).

Lichtenstein, W. A. J. Stylopized specimens of solitary and social wasps from Mexico (1886).

Lidderdale, Dr *see* Doncaster, L. (1927).

Lister, J. J. British, Palaearctic and Ethiopian insects, mainly Lepidoptera (presented by B. G. Adams, 1928).

LIVESEY, R.   South American Coleoptera, Hemiptera, Orthoptera, etc. (out of cabinet purchased at Stevens's saleroom, 1923).

LIVINGSTONE, DR DAVID   A card bearing three fragmented specimens of *Glossina morsitans* marked in Westwood's handwriting 'Dr. Livingstone' and in pencil 'Setse' (n.d.).

'Leaves of a species of *Bauhinia* from Central Africa coated with exudation of a Psylla collected & eaten by the natives received from Dr. Livingstone'; also female and young of *Argas* 'which enter into the feet of negroes of Central Africa between the toes causing inflammation & other diseases', two larvae of a Chrysomelid beetle 'used by the native africans when crushed for poisoning their arrows', and a new species of 'Phyllomorpha' from Central Africa (1 September 1857).

LLOYD, R. W.   Many small donations from British and European localities (1901–55).

LOAT, W. L. S.   Butterflies from near Kaka and Gharb-el-Aish on the White Nile (1901).

Series of insects, chiefly butterflies, from the reaches of the White Nile, northern boundary of Uganda, also a collection of insects of various orders from the Blue Nile (1902).

LONGSTAFF, G. B.   Insects collected from world-wide localities (1904–14).

Collection of insects from Mortehoe, North Devon (1906, 1908, 1912).

Further collection of Lepidoptera from Mortehoe, also insects taken on yearly journeys, books and atlases (1921). Many of the specimens figured in published work.

Fine old Oertling balance, his prize from St Thomas's Hospital, 1874.

LOPEZ DE SEOANE, V.   A few Spanish Orthoptera to Westwood (1879).

LORENA, DR and MRS A. C.   Butterflies from the Gold Coast (1915–22).

LOVERIDGE, A.   Coleoptera from German house, Cameroons, 1914.

The material illustrating a raid by 'Siafu' (Driver ants) at Kilosa, Tanganyika Territory, East Africa (1922).

LOWE, REVD F. E.   European Lepidoptera *via* H. Rowland-Brown (1922).

LUBBOCK, J.   'Pauropus (minute G Myriapoda 2 ind & skin) Mymar natans & Buskii (several indiv⁹)' (to J. O. Westwood, 1866).

Four syntypes of *Polynema natans* Lubbock in Hope Collections (see Graham, M. W. R. de V., 1983. *Proc. R. Irish Acad.* (B) **82 B** (12), pp. 218–19). Also one syntype of *Prestwichia aquatica* Lubbock; this may be unique as no others have been found in the Lubbock collection in the British Museum (Natural History).

LUCAS, W. J.   Series of insects of all orders from British localities (1900, 1901, 1902).

Neuroptera and Orthoptera from world-wide localities (1904).

British Odonata, Neuroptera, Dermaptera, and Orthoptera (1932). Collection to remain intact. He also presented many other insects for the British Collections and bionomic series during his lifetime.

LUPTON, MISS P. M. and WYKES, MISS U. M.   About 370 insects of various orders from south-west Iceland (1935). Dried material, microslides and one jar of specimens in spirit.

MACALISTER, A.   Two cases of *Oiketicus* (two n.sp.) from New Holland (presented by J. O. Westwood, 1867).

MACAN, T. T.   Collection of Ephemeroptera nymphs and adults preserved in formalin (1968).

McArthur, H. *see* Doncaster, L. (1927).

McComish, Mrs I. Small collection of insects from Norfolk Island 1939 (*c*.1940) (see Hawkins, C. N. 1942. *Ann. Mag. Nat. Hist.* (11) **9,** 865–902). Material appears to have been presented to the British Museum (Natural History) and has the Catalogue No. BM 1940-154.

Mackinder, H. J. and Hausburg, C. B. Insects of all orders from British East Africa including some species new to science (1899).

McLachlan, R. 152 specimens (39 species) of British Trichoptera (1862).
 Bittacus chlorostigma and other exotic Panorpidae (1886).
 Collection of dragonflies from Java (1900).

Macleay, W. S. Eight Australian Coleoptera, including a large black species of *Clerida*; and a ♂ and ♀ *Thynnus* (in exchange for *Xenos vesparum*, 1862).
 Many insects sent to F. W. Hope (*c*.1842).

McLeod, Sir M. 86 specimens of Lepidoptera, many from Oberthür, from tropical America and Asia (1943).

Macquart, P. J. M. Diptera in part *via* Bigot to Verrall-Collin collection (1967).

Malaise, R. 45 Tenthredinidae, including 13 paratypes, from various localities (by exchange, 1933).
 Four species of *Tenthredo,* three of which are paratypes, from north-east Burma (1935).

Manders, Col. N. Butterflies mainly from Curipepe and a few from other localities in Mauritius, including a series of the Satyrine butterfly *Melanitis leda* showing seasonal variation, also seven butterflies from Bourbon, Réunion Island (1907).

Mannerheim, Comte C. G. von 60 species of Coleoptera to Hope (June 1830).

Marchal, [G.] Collection of Orthoptera and other insects *via* F. W. Hope (n.d).

Mark, E. W. 245 Lepidoptera collected mostly in the neighbourhood of Bogotà between 1848 and 1857, also a large number of other insects (presented by F. W. Mark, 1901).

Marley, H. W. Bell *see* Bell Marley, H. W.

Marshall, Maj. C. C. *see* Wilson, Capt. R. S. (1915–22).

Marshall, G. A. K. Insects of all orders, including many butterflies, also bionomic material from South Africa and Southern Rhodesia (1896–1906).
 Collection of Orthoptera, chiefly Acridiidae, from Salisbury, Mashonaland (identified by Bolivar), also mimetic insects of various orders from same locality with large numbers of the Hymenoptera which form the models most commonly resembled (1900).
 Hymenoptera and Diptera from Wroxham, Norfolk (1902).
 Series of Types and syntypes of Curculionidae (described in *Proc. Zool. Soc. Lond.* **1906,** p. 911) (1907).
 Collection of butterflies from Trinidad (1912).

Marshall, Revd T. A. Hemiptera from world-wide localities (presented by E. Saunders, 1905). Some specimens may have been labelled erroneously J. A. Marshall.

Martius, C. F. Ph. de Insects to Hope obtained from Gistl and Döbner (1833).

Mason, C. 640 butterflies from Nyasaland (1938).

Mason, J. Wood- *see* Wood-Mason, J.

Mathew, G. F. Diurnal Lepidoptera (from sale, April 1889, £2. 9*s*.).

MATTHEWS, REVD A. Collection of named British Coleoptera Trichopterygidae (1886).

MAVROMOUSTAKIS, G. A. Insects from Cyprus (by exchange 1929–51).

MEADE, R. H. 102 species of British spiders 'carefully named by Mr. Meade' (lot no. 110, July 1863 at sale of Entomological Society's British Collection, 11s.).
For British Diptera, *see* WALKER, F.

MELANDER, A. L. *see* VERRALL, G. H. and COLLIN, J. E.

MELDOLA, R. Collection of butterflies from Trinidad and insects of all orders from Canada (1897).
Insects, chiefly Lepidoptera, from Switzerland and a few from Cap Gris Nez (1899).
Collection of Lepidoptera from Trinidad (1901).
British Lepidoptera and many books (1915).
Mrs Meldola also presented many publications to the library during her lifetime.

MERFIELD, F. G. *see* POWELL-COTTON MUSEUM (1937, 1938).

MERRIFIELD, F. Butterflies bred from European larvae (1900).
Bred specimens of the butterfly *Araschnia levana* with pupa cases (1911).
Lepidoptera, including his temperature-experiment material (1924).

MEYER, E. L. Insects from Schleswig, Upper Bavaria and Munich (1902) and over 2000 from the Malayan region (1904).

MICHAELIS, H. N. 56 British Micro-Lepidoptera (1959); 126 specimens of the Tortricid *Peronea* (1960).

MICHELMORE, A. P. G. *see* CAMBRIDGE UNIVERSITY EXPENDITIONS.

MIERS, J. 30 honeybees and 14 honey-making wasps from South America which were exhibited at the International Exhibition, Brazilian section, in 1862 (1865).
Insects, mainly beetles, comprising over 37 000 specimens principally from South America (presented by his son J. W. Miers, 1880). His Chile collections were lost at sea.

MILLAR, A. D. Series of butterflies bred from a female *Hypolimnas* (*Euralia*) *mima* captured near Durban, together with pupa cases, also a bred series of *Pseudacraea* and other Natal butterflies (1910).

MILLAR, H. M. and BLAKE, J. Lepidoptera from various parts of the world, and a coloured drawing of a remarkable specimen of *Papilio pylades moraria*, the original butterfly having been presented to the department in 1932 by Mr Millar (1955).

MINCHIN, E. A. Selection of rare Indian Lepidoptera with illustrations of the habits of many of them (July 1886, £5).
Indian Lepidoptera Rhopalocera and Coleoptera (1889, 1893).
63 Lepidoptera from Entebbe and Kampala (1905).
See also LANCHESTER, W. F. and MINCHIN, E. A. (1909).

MJÖBERG, E. *see* HANITSCH, K. R.

MNISZECH, COMTE G. DE Coleoptera from Lake N'Gami, Caffraria, Abyssinia, Malacca, Sumatra, China, etc. (by exchange, 1862).

MODIGLIANI, E. *see* HANITSCH, K. R.

MOGGRIDGE, J. T. Series of spiders' nests (*c.*1873).

MONTEIRO, J. J. 88 butterflies, 13 moths, and 14 beetles from Angola (December 1874, £6. 17s. 9d.).

MOORE, F. Specimens of native Indian silk and cocoons (1863).

Pupa of the genus *Epicopeia* Westwood (1864).
Fragments of malt liquor cask from India infested by *Tomicus* sp. (1867).

MOORE, J. W.   Small series of butterflies and moths from Peru and western China (1944).

MOORE, M. S.   Insects from Tanganyika, chiefly Asilids with prey (1937).

MORICE, REVD F. D.   Seven Aculeate Hymenoptera from the flowers of a single tree of *Ochrademus baccatus*, in the dry bed of the Wady Kelt near Jericho (1909).
15 *Pteronidea spiraeae*, a sawfly new to Britain (1925).
Large collection of Palaearctic Hymenoptera, especially Tenthredinidae and Aculeata, together with books and offprints (1926). Collection contains material from Friese, Schmiedeknecht, Konow, Pérez, Du Buysson, and also some from P. A. Buxton.

MORRELL, R. D'A.   Coleoptera of various groups from Virginia, United States (1900).

MOSELEY, H. N.   Specimens of galls and various Lepidopterous insects from Barbary (1886).

MOSSE-ROBINSON, L. H.   Set of Australian insects, chiefly moths, also Lepidoptera from Fiji, Ceylon, etc. (1913).

MOTSCHULSKY, V.   Collection of Trichoptiliens (Coleoptera, Trichopterygidae) collected by Motschulsky (received from M. Westermann of Copenhagen and presented by J. O. Westwood, 1858).

MOULTON, J. C.   13 butterflies from Sarawak (1912).
Six males of the Pierine butterfly *Delias eumolpe* from Mount Kinabalu, British North Borneo (1913).
Malayan butterflies (1915–22).

MOUND, L. A.   23 microscope slides of British and foreign Thysanoptera (1966).

MOUNT EVEREST COMMITTEE   *see* BULLOCK, G. H.

MOYSEY, MAJ. F.   41 butterflies from Mount Elgon, Uganda (1932).
447 butterflies from neighbourhood of the Sudan–Uganda–Kenya frontier (1936).

MUIR, F.   Insects of various orders from Natal illustrating bionomic principles and mimicry (1903).
See also KERSHAW, J. C. (1909).

MURRAY, A.   Coleoptera Heteromera from Pacific side of North America and Cleridae from various localities (lots 53 and 166, from sale of his collection by auction April 1878, £2. 13*s*.).

MUSÉE NATIONAL D'HISTOIRE NATURELLE, PARIS   Duplicate Blattidae *via* R. Shelford, including syntypes (1906–7).
Series of Orthoptera, Blattidae, and Phasmidae (1911).

MYERS, J. G.   Insects from Venezuela, British Guiana, Trinidad, Haiti, and South Sudan (1930, 1931, 1935, 1940).

NATURAL HISTORY MUSEUM, LUXEMBOURG   69 butterflies from Kondué, Kasai, Belgian Congo (1925).

NAVAS, L.   Insects to Poulton (1913).

NEAVE, S. A.   Insects from Rhodesia, also a large and varied collection made in 1904–5 when he was naturalist to the British South African Company (1904–9).
Collection of 789 butterflies from south-east Congo Free State and 98 Blattidae collected in the Congo Basin (presented by Musée du Congo, Brussels, 1909).

Series of butterflies from British New Guinea, nearly all collected by A. E. Pratt, by exchange with G. T. Bethune-Baker (1910).

Butterflies from Madagascar, Kilimanjaro, and German East Africa acquired by exchange with naturalists and museums (1911).

14 Asilid flies and Xylocopid bee models from Mount Mlanje, Nyasaland, 1913, and a series of *Pseudacraea dolomena albostriata* and its form *dolabella* from Uganda, 1911–12 (1925)

See also COLONIAL OFFICE.

NEES VON ESENBECK, C. G. *see* ESENBECK, C. G. NEES VON.

NEVINSON, E. B. 7000 British Aculeate Hymenoptera and 556 dragonflies (presented by R. C. L. Perkins, 1938).

'The E. B. Nevinson Collection of British Chrysididae presented by Willoughby Gardner, Hon. D.Sc. (Wales), F.L.S.' (1952). Described and acknowledged as donor wished.

NEWMAN, E. Part of his collection *via* F. W. Hope; contains a few Types of sawflies and Chalcidoidea.

NEWNHAM COLLEGE, CAMBRIDGE *see* HAVILAND, MISS M. D.

NEWPORT, G. Myriapoda, including some Types (n.d.)
See also SMITH, F.

NEWSTEAD, R. Set of Coccidae collected and named by donor (May 1894).

NIBLETT, M. Bred series of about 40 species of British Cynipoidea (1958).

NICÉVILLE, L. DE Series of two forms of a Pierine butterfly, *Catopsilia crocale* (8) and *C. catilla* (6) and a pair of another species *C. pyranthe* flying together in the Kangra Valley, west Himalayas (1900).

NIETNER, J. Specimens of *Myrmecolax Nietneri* Westwood (Strepsiptera) and the ant infested by it (1861).

NORRIS, T. Purchased at sale of his collection, lots 206 and 222 (May 1873, £2. 15*s*. 6*d*.). Lot 222 comprised the whole of his *Castnia*.

NURSE, COL. C. G. Asilidae from Karlsbad, Mentone, Simla, Punjab, Bombay, and some British localities (1902).
Series of Oriental Asilid flies (1907).
Some Aculeates *via* Rothney Collection (1920).

NUTTALL, G. H. F. Small collection of Ixodoidea presented to E. B. Poulton (n.d.).

OATES, F. 'Collection of Caffrarian insects formed by Mr. F. Oates' from 1873 to 1874. (*See* Appendix by J. O. Westwood to *Matabele Land and the Victoria Falls*, edited by C. G. Oates, London 1881.)

OBERT, J. Coleoptera from Siberia to Hope (1842).

OBERTHÜR, C. Examples of rare mimetic Lepidoptera from China and Sikkim (1904).
Series of butterflies of the genus *Neptis* from Madagascar and surrounding islands (1908).

31 Acraeine butterflies collected 1883–96 from various localities in Madagascar, chiefly by the Perrot Brothers (presented by S. A. Neave 1911).

122 Acraeinae from Madagascar (1912).

Butterflies from China, Java, Madagascar and the Comoro Islands, Congo State, and Cameroons (1913).

See also MCLEOD, SIR M. (1943).

OBERTHÜR, R. Various rare exotic Coleoptera (1880, 1886).

OCSKAY VON OCSKO, BARON F.   Orthoptera from Hungary and adjacent provinces to Hope (1835 and 1837).

OGILVIE-GRANT, W. R.   94 butterflies from Socotra and one from south-west Arabia (1900).

OLLENBACH, O. C.   Series of butterflies from United Provinces, northern India (1915–22, 1925).

O'MAHONY, E.   23 slides of fleas from Ireland (through R. B. Freeman 1940).
Slide of two females of *Damalina bovis* (Mallophaga) from Cornwall (1947).
Collection of 644 slide mounts of fleas, 585 of lice, also 12 209 Irish Coleoptera and 322 insects of other orders (presented by his family in accordance with his wishes 1951). Some material in spirit collection.

ORMEROD, MISS E. A.   Coleoptera from Madeira named by Mr Bewicke (1860).
118 butterflies from Penang and 16 dragonflies (1862).
Various species of insects injurious to agricultural and horticultural produce (1885).
Horse bot flies and ox hide with perforations by warbles and photographs of same (1888).
Pulvinaria ribesiae and specimens of two species of Hippoboscidae parasitic on the Ostrich and Cape Pigeon (1889).
Plum-tree shoots injured by burrowing beetle *Tomicus dispar* (1890).
Five Hippoboscidae from Ross-shire sent to Dr W. Hatchett Jackson, 1896, package found unopened 1935 (1939).

ORMISTON, W.   Butterflies from Ceylon (1915–22, 1925).

OSMASTON, B. B.   11 male butterflies, part of a crowd drinking at mud on the road between Kampala and Mubende, Uganda, and an unusually large male of *Acraea anacreon* from near Cape Town (1927).
30 butterflies from Kashmir, Ladak, and Baltistan (1928).
21 insects captured on the wing in trios, each trio made up of a paired ♂ and ♀, and another insect presented probably as a wedding gift by the male, from Morogoro, Tanganyika Territory (1929).
Coleoptera and some Hemiptera from the Himalayas, 1889–1900 (n.d.).

OVERBECK, H.   Collection of Blattidae from Trial Bay, New South Wales, made whilst prisoner-of-war 1914–19 (presented by K. R. Hanitsch, 1926).

OWEN, CAPT. I. G.   Over 13 000 butterflies from southern Sudan and Congo–Nile watershed (presented during his service in the Sudan, 1933–7).

OXFORD UNIVERSITY EXPEDITIONS   Material in both dry and spirit collections.
Spitsbergen 1921, 1923 (Merton College), 1924.
Greenland 1928.
Lapland 1930.
Hudson Strait 1931. Akpatok Island, Ungava Bay.
Borneo 1932. Collection to British Museum (Natural History), duplicates to Hope Department and Sarawak Museum. 28 Trichoptera, mainly paratypes (presented by British Museum (Natural History) 1954).
Arctic 1933. West Spitsbergen.
Ellesmere Land 1934–5.
Arctic 1935–6. North East Land.
Greenland 1936.
Cayman Islands 1938. Insects retained in Hope Department until February 1967 when they were transferred to the British Museum (Natural History). Some Types,

however, remain in the Hope Collections. In 1941 the British Museum (Natural History) donated 83 Hesperiidae.

Lapland 1938, Womens' Expedition.
Jan Mayen 1947.
Tanganyika 1958.
Seychelles 1975, 1978.
Spitsbergen (Svalbard) 1978.

PAIN, T.   143 butterflies and two moths from British Guiana (purchased 1948).

PALLIS, M. A.   Coleoptera from West Africa, *c.*1906 (n.d.).
Coleoptera from Brazil, Trinidad, and British Guiana, *c.*1911–13 (n.d.).
Lepidoptera from many localities (1915–22).

PALMER, M. G.   *see* DRUCE, H. (1910–12).

PARKER, D. E.   739 insects of many orders reared from elm logs during investigations by Parker and J. M. Walter on behalf of the US Department of Agriculture into Dutch elm disease (1938).

PARKINSON, F. B.   Insects, mainly Coleoptera, Hymenoptera, and Orthoptera, also Arachnida and Acari from Baviaan Krantz, Orange River Colony (1906).

PARMENTER, CAPT. L.   83 specimens of Asilidae, Empididae, and Cordyluridae, with prey, from Cornwall, Sussex, and Surrey (1937).
105 various predators and prey (1941).
94 Asilids and one *Empis* with prey (1942).

PARREYSS, L.   Coleoptera purchased by Hope (*c.*1834–6).

PARRY, MAJOR F. J. S.   Insects from China, Malacca, Singapore, and Natal (by exchange, 1858).
Insects (1860).
Ornithoptera brookiana and *Lucanus reichii* (by exchange, 1861).
Insects from Ceylon, Cambodia, and West Coast of Africa (1862).
Few specimens of *Lucanus* and *Dorcas* from Borneo and West Africa and Coleoptera from Hyères (by exchange, 1863).
Some *Corynophyllus*, *Polyommatus*, and *Odonestis* from southern Spain (1867).
Rare Lucanidae and Cetoniidae (from Parry sale, 1885, £10. 12*s.*).

PARSONS, G. L.   Insects from South Africa, and a few from Beira and Mombasa (1905).

PASCOE, F. P.   Coleoptera, mainly from north China (by exchange, May 1858).
Coleoptera from Hyères (1863).
The major portion of his collection, including Types, went to the British Museum (Natural History) in 1893; the remainder, occupying 13 cabinets, and part of his library, was presented to Hope Department by Miss Pascoe at the suggestion of A. Russel Wallace (1909).

PASSERINI, C.   Insects to Hope by exchange (*c.*1834–53). Insects to Westwood (n.d.).

PEACHEY, MRS   *see* RUSSELL, C. B. (1902).

PEARCE, MISS E. K.   British Diptera and 16 boxes of negatives (1940). Includes specimens illustrating her work *Typical Flies*, 1915, series II, 1921, and series III, 1928.

PECCHIOLI, V.   Italian insects to Hope by exchange (*c.*1834–44).

PEEL, C. V. A.   Two specimens of *Limnas Klugii* captured in Somaliland, and a series of *Catocala promissa* from Lyndhurst (1896).
Collection of insects and a few Myriapoda and Arachnida from Somaliland

(1897). Contains Types or syntypes described by donor and others in *Proc. Zool. Soc. Lond.*, **1900**.

Insects, mainly butterflies, from British Central Africa and south-east Africa (1900).

127 butterflies and one moth from North Providence Island, Bahamas (1902).

PEILE, COL. H. D.   Collection of carefully selected butterflies, many rare, from the Himalayas (1937).

817 butterflies from North-West Frontier of India (1942).

Large collection of butterflies from India, Iraq, north-west Persia, and a few from Spain (1948-9). The bulk of this collection, which includes many rare forms, is still unset.

PELTONEN, E. O. D.   *see* WOOLLATT, L. H.

PEMBERTON, C. H.   Insects of all orders from British Central Africa, also a small collection of butterflies made at Delagoa Bay (presented by C. V. A. Peel, 1900).

PENDLEBURY, H. M.   *see* HANITSCH, K. R.

PÉREZ, J.   *see* MORICE, REVD F. D.

PERKINS, R. C. L.   Two Diptera, with their two Hymenopteran models, from Arizona (1898).

A pair of the very rare Hymenopteron *Astata stigma* from Brandon, Suffolk (1899).

Series of butterflies collected mainly in southern Arizona, and a pair of *Prosopis palustris*, a bee new to science, from Wicken Fen, Cambridgeshire (1900).

Butterflies from Queensland collected in 1904 (1911).

Series of Hawaiian wasps (1912, 1927).

Fine collection of Hawaiian insects, and bred series of *Pararge aegeria* demonstrating difference between two early broods of this species in England (1915-22).

British Aculeates, Chrysididae, and sawflies, including F. Smith's collection of Aculeates, also a series of Aculeates from the Hawaiian Islands (1930) with further additions (1931, 1942-3, and 1946). The collection contains most of the Arnold material which was purchased by Perkins.

PERKINS, V. R.   Four specimens of a new British bee, *Andrena angustior*, from Wootton-under-Edge (1891).

PERROT BROTHERS   *see* OBERTHÜR, C. (1911).

PETERS, C. J.   Bird-catching spider from India (1888).

PICKARD-CAMBRIDGE, SIR A. W.   Collection of insects made by him and his father O. Pickard-Cambridge, includes a wide variety of Lepidoptera and Coleoptera and collection of Micro-Lepidoptera and Hemiptera (1952).

PICKARD-CAMBRIDGE, F. O.   Some material collected by F. O. Pickard-Cambridge, nephew of O. Pickard-Cambridge, is included in the latter's collection.

PICKARD-CAMBRIDGE, REVD O.   Collection of Arachnida and library on Arachnological subjects. About 5000 bottles of all sizes, many holding from two to 20 separate tubes, comprising an almost complete British collection, and a very large collection of exotic species from every part of the world. Contains many hundred Type specimens, not only those described by him but also those of John Blackwall and a considerable number from eminent specialists such as Simon, Koch, and Thorell. The library consisted of about 140 bound volumes and a large number of pamphlets and offprints. Poulton described the bequest as 'the greatest contribution to systematic zoology that the University had ever received by one gift' (1917).

Previous donations: seven bottles of objects in spirit, chiefly fishes and reptiles from Egypt (1865); six rare Tortricidae from Dorset (1891); set of insects, chiefly Lepidoptera, from the Potaro River, Central British Guiana (1897).

For insects collected in Palestine in 1865, *see* TRISTRAM, CANON H. B. (1897).

PILCHER, MAJ.-GEN. J. G.   Lepidoptera from Sikkim (1894).

Large collection of North Indian moths (presented by Mrs Pilcher, 1917).

See also DONCASTER, L. (1927).

PITMAN, CAPT. C. R. S.   Butterflies from Kenya and Uganda collected 1923–4 and *c.*1928 (n.d.).

♀ Type and paratype of new Blattid *Ischnoptera pitmani* from Entebbe (1928).

Series of the Nymphaline *Cymothoë caenis* migrating in Uganda (1934).

PLATT, E. E.   Bred specimens of butterflies from Durban district (1914).

Seven *Charaxes xiphares* from Natal (1938).

PLATTS, MRS H. C.   Lepidoptera from Pemba Island (1927).

PLAYNE, H. C.   13 *Vanessa urticae* var. *ichnusa* and one *V. atalanta* from Corsica (1897).

PLAYNE, H. C. and WOLLASTON, A. F. R.   Set of insects from north-west Finland (1897).

POGSON SMITH, W. G.   *see* SMITH, W. G. POGSON.

POLLARD, F. E.   Butterflies captured in 1894 at Etshowe, Zululand (presented by Miss Pollard, 1895).

POMEROY, A. W. J.   Lepidoptera from the Gold Coast (1927, 1928, 1930).

Lepidoptera, mainly moths, and four mounts of Diptera Phoridae, from Kingston, Jamaica (1932).

POMEROY, W. K.   57 butterflies of several groups from Colombia (1914).

PONT, A. C.   Small collection of Diptera, mainly from Scandinavia (1963).

296 specimens (24 species) of Muscidae from Swedish Lapland (1966).

POOLE, C. J. C.   British collection (June 1912, £15).

POPPLEWELL, H. B.   71 butterflies and two moths from British East Africa (1913).

PORTER, W. H.   Insects sent from New South Wales to Hope (1834).

POULTON, PROF. E. B.   Pupae of *Vanessa urticae* showing various colours (four specimens, two gilded) (1887).

Lepidoptera from Tenerife, Grand Canary, and Madeira (1893).

Many donations through collecting and purchase during his tenure of the Hope Chair, from Britain and various parts of the world, including Switzerland, Majorca, Minorca, Norway, and Iceland, also from Canada and United States where he made a tour and a vast amount of material collected by him and his associates was donated.

Books and separata from his library (presented by his daughter Mrs Maxwell Garnett, 1944).

POWELL-COTTON MUSEUM   1823 butterflies from Lomie district, Cameroons (purchased 1937, 1938); collector F. G. Merfield.

PRANGNELL, C. F.   A hundred specimens of the Staphylinid beetle *Deleaster dichrous* found in damaged potato (1889).

PRATT, A. E., C. B., and F. B.   Butterflies from Tibet and western China (from A. E. Pratt, 1893, £7. 4*s.* 4*d.*).

Lepidoptera from Peru, Madagascar, and New Guinea (presented by J. J. Joicey, 1912, 1913).

See also NEAVE, S. A. (1910), DONCASTER, L. (1927), and DRUCE, H. (1910–12).

PRATT, H. C.   A few Blattids from Dutch New Guinea (1906).
Series of Orthoptera, chiefly Acridiidae, from Selangor, Malay Peninsula (1908).

PREST, E. E. B.   161 rare moths much needed in British collection (purchased December 1923, £5. 10s.).

PROUT, L. B.   Material used in the Mendelian breeding experiment on *Acidalia virgularia* (5000–6000 specimens) with A. Bacot (1909).

PRYER, H. J. S.   *see* DONCASTER, L. (1927).

PRYER, W. B.   Fine collection of butterflies made in British North Borneo, 1878–98 (presented by Mrs Pryer, 1900).

PULS, J. C.   Beetles destructive to species of *Zamia* in Australia and South Africa (1885).

PUSEY, LIEUT.   A number of insects, dried and in spirit, from Vancouver Island (1864).

PUTZEYS, J. A. A. H.   Some Types of *Clivina* (Col. Carabidae) in Hope's collection, also in Chevrolat collection; see manuscript note (ANDREWES, H. E., 1918, in Appendix A).
See also BADEN, F. and SOMMER, M. C.

RADDON, W.   Cabinet *via* Hope Deed of Gift (1849).
Hymenoptera and Coleoptera from turpentine sent from North America *via* J. O. Westwood (1857).

RAMBUR, J. P.   Odonata, Neuroptera, and a few Isoptera Types *via* Marchal Collection. Some Odonta destroyed or damaged through accident in department.

RAMME, W.   Large collection of Blattids from north-east New Guinea collected by Dr S. G. Bürgers during the Kaiserin Augustafluss expedition (1934).

RATTRAY, MAJ.   *see* WIGGINS, C. A. (1903).

RAWNSLEY, LIEUT. A. E.   Trapdoor spiders and nests from Jamaica (1888).

RAYNOR, REVD G. H.   Series of *Abraxas grossulariata*, bred and named by donor (1929).

REED, E. C.   780 specimens selected from his Chilean collection, mainly Hymenoptera, Diptera, Lepidoptera, and Hemiptera (April 1874, £10).

RÉGIMBART, M. A.   *see* ANDREWES, H. E. and F. W.

REHN, J. A. G.   Series of Orthoptera, chiefly Blattidae, determined by donor (1907).
See also BURR, M. (1925).

REICH, G. C.   Insects to Hope by exchange (*c.*1833–45).

REICHE, L.   Insects to Hope (1844).

REID, E. T. M.   A few butterflies from Nyasaland (1948).

RENGARTEN, K.   Collection of Coleoptera from the Klouchi goldmine, in the neighbourhood of Gorbitza, left bank of the Shilka River, Transbaikalia, preserved in vodka (presented by D. P. Lance, 1905).

RICHARDS, O. W.   Series of Asilidae with insect prey from various British localities including Scilly Islands, also Europe and North America (1931).
74 specimens of Hymenoptera with prey (1937).
Two specimens of the Sphecid wasp *Ectemnius* (*Clytochrysus*) *zonatus* and a series of Syrphid prey from nests in timber in a quarry, Alderney, Channel Islands (1958).

RICHARDS, W. S.   39 slides illustrating the life-history of the mite *Trombicula autumnalis* (presented by Professor P. A. Buxton, 1949).

RICHARDSON, H.   Small series of Lepidoptera, mostly butterflies, from Switzerland (1899).
Butterflies from Ospedaletti, near San Remo (1901).

RICHARDSON, L.   A few butterflies from Mexico and Trinidad (1898).

RICKARDS, A.   Insects from Van Diemen's Land (May 1858).

RIDLEY, H. N.   138 insects captured in the Botanic Gardens, Singapore (presented by R. Shelford, 1906).
Insects of various orders (1906–8, 1912).

RIGAUD, CAPT.   Harlequin beetle, gigantic *Scolopendra* and several scorpions from Barbados (1888).
Two nests of a Brazilian *Oiketicus* and a young *Mygale* (1889).

RIJKSMUSEUM VAN NATUURLIJKE HISTORIE, LEIDEN   Indian Diptera, Hemiptera, and Hymenoptera to Hope through W. de Haan (n.d.).
Coleoptera from Japan, Java, Caffraria, Timor, and Borneo to Hope through W. de Haan (1830–1).
Insects from Japan, Java, Bogotà, Borneo, Celebes, and Sumatra from Snellen van Vollenhoven (1858).
Insects from Sumatra, mainly Coleoptera (1861).
Specimens of all orders from Oriental Region and Caffraria (1863) and Borneo and the Malay Archipelago (1864).
A few beetles from New Guinea and Malay Archipelago (1865).

RILEY, C. V.   Male and female *Hornia*, a wingless beetle (1888).

RITSEMA, C.   *Gymnopleurus* from Liberia (1887).
A species of the Homopterous genus *Derbe* (1889).

ROBERT, C.   Insects from the environs of Liége to Hope (1836).

ROBERTS, C. B.   323 butterflies collected near the Potaro River, British Guiana, all the specimens having been captured on the white flowers of *Eupatorium macrophyllum* (1903, £3. 5s.). Another set taken in the same locality (1904, £3).

ROBERTSON, MAJ. R. B.   Insects of many orders, mainly from Hampshire (1902, 1906).

ROBINSON, A.   Preserved caterpillars of several species of British moths (1890).

ROBINSON, H. C.   Series of eight males and 10 females of the butterfly *Ornithoptera poseidon euphorion* in part bred and in part captured at Cooktown, north Queensland coast (1903).
Butterflies from the Federated Malay States injured by the attacks of enemies (1915–22).
See also ANNANDALE, N. (1903).

ROBINSON, L. H. MOSSE-   *see* MOSSE-ROBINSON, L. H.

ROBSON, S.   *see* DONCASTER, L. (1927).

ROE, J. E.   Insects from Swan River, Western Australia to Hope (1833, 1836, and 1844).

ROE, W.   Insects from world-wide localities to Hope (1839).

ROGERS, LIEUT. G.   Butterflies from Andaman and Nicobar Islands, 1903–4 (1906).

ROGERS, H.   Lepidoptera from Rogers Collection (June 1873, £3. 19s.).
Lepidoptera from Gaboon (October 1873, £3. 6s.).

Coleoptera from Angola (October 1873, 17*s.* 6*d.*).

Lepidoptera (July 1874, 6*s.* and 6*s.* 6*d.*).

All these were purchased from Mr Hewitson (*see* HEWITSON, W. C.).

ROGERS, K. ST AUBYN   Lepidoptera, chiefly butterflies, from British East Africa, Kenya, Tanganyika, South Africa, and Trinidad, and extensive donations over many years (1904–35).

ROHU, H. S.   325 Lepidoptera and three Homoptera from British New Guinea (1902, £1. 5*s.*).

Insects of various orders from New Guinea (presented by H. Balfour, 1907).

ROLLE, H.   Hymenoptera from Africa, Madagascar, Madeira, Formosa, China, Java, India, Ceylon, etc. (presented by G. A. J. Rothney, 1911) and miscellaneous bees from Turkestan and Himalayas (n.d.).

RONALDS, A.   Original specimens used to illustrate *The Fly Fisher's Entomology*, 1836, *via* J. O. Westwood (1857). (Kept separate.) Arranged by M. E. Moseley—recent specimens stand side by side with the originals, many of which are reduced to fragments.

ROSE, CAPT. H., 'SHIP MERMAID'   'Elephant fish' caught in Lyttleton Harbour, Canterbury, New Zealand, not full grown, and a pipe fish (1864).

ROSENBERG, W. F. H.   293 butterflies, moths, and a few other insects from Ecuador, many of great value for the Mimicry Collection (1899, £9. 10*s.*).

199 butterflies from Kikuyu Escarpment, East African Protectorate, collected by the late W. Doherty, and 150 Lepidoptera from La Merced, Peru, and Upper Rio Toro in same province, many exhibiting mimetic resemblances and some cryptic characters (1902, £30).

82 insects, chiefly Coleoptera, from British East Africa and Ecuador (presented 1903). Western tropical American butterflies (1903, £2. 2*s.*).

144 Lepidoptera from Chubut in the Patagonian Andes (1904, £2. 10*s.*).

Seven specimens of the Danaine butterfly *Amauris* from Bitja, Ja River, Cameroons (1912, 16*s.* 1*d.*).

Five Papilios from Formosa (1913, 6*s.* 6*d.*).

Four mimics (Hymenoptera and Coleoptera) (1913, 7*s.* 6*d.*).

ROSS, E. S.   44 butterflies from Maffin Bay, Dutch New Guinea (1948).

24 interesting species of Embioptera from Congo, Kenya, Colombia, and Mexico (1960).

ROSS, SIR J.   Some Types in Dale collection of Lepidoptera described by J. Curtis in *Appendix to the narrative of a second voyage in search of a north-west passage, 1829–33*. Actually published 1835.

ROTHNEY, G. A. J.   Fig insects from Brisbane, New South Wales (1887).

Collection of oriental Hymenoptera and British Aculeate Hymenoptera, also oriental insects of various orders and some South American Hymenoptera, library and manuscript notes (1910). The oriental Hymenoptera contain material collected by C. T. Bingham and specimens from collections of Bates, Cameron, Forel, and F. Smith, and is rich in Types. This gift also included some butterflies and dragonflies from Barrackpore Park.

Three important collections—Hymenoptera from Africa, Madagascar, Madeira, etc., from Hermann Rolle; Malayan Hymenoptera from the Van der Poll Collection; and a collection, chiefly Hymenoptera, made by F. P. Dodd in north Australia. (1911).

Further donations, chiefly Aculeate Hymenoptera, from India, Australia, West Africa, and North and South America, also Hymenoptera from the collections of

E. Saunders, F. Smith, and F. Walker (1912, 1915–22). The Indian series includes 1320 Aculeates (621 bees) from Colonel C. G. Nurse's collection, selected by Professor T. D. A. Cockerell, and purchased 1920 by Rothney for £7. 10s.
Lepidoptera from Darjeeling and Calcutta (1925).

ROTHSCHILD, HON. [L.] W.   Lepidoptera and a few Homoptera from the forests of the Luebo district, Kassai River, Congo Free State (1906).
Lepidoptera from German New Guinea and New Ireland, oriental Papilionidae, and Longicorn beetles from the Ethiopian region (1909).
43 butterflies from the Gold Coast (1910).
One specimen of *Papilio laglazei*, mimic of the Papuan moth *Alcidis aurora*, from Dutch New Guinea, and 13 examples of *Papilio memnon agenor* from Loo Choo Islands (1911).
Acraeine butterflies (1912).
Pair of *Ornithoptera alexandrae* bred from larvae found in the Owen Stanley Range, south-east New Guinea (1913).
Various small donations (1918–35).

ROWLAND-BROWN, H.   European butterflies (1914).
European Lepidoptera, including collection of the late Revd F. E. Lowe, together with a large number of books (1922). The collection contains eight specimens of the extinct 'Large Copper' butterfly. A fine microscope which previously belonged to Sir W. S. Gilbert was presented by Miss Rowland-Brown in accordance with the known wish of her brother.

ROWLEY, REVD H.   Coleoptera and various other insects collected near River Zaire while he was atttached to the Oxford and Cambridge Mission to the Zambesi (1864).
Collection of insects from Zambesi (1864, £9).

RUSSELL, C. B.   Butterflies from near Eshowe, Zululand (1900).
199 butterflies and one moth from Llabisa, East Central Zululand, collected by Mrs Peachey (1902).
Few insects from Kenya, Uganda, and Brazil (1942).

ST FARGEAU, COMTE DE   *see* LEPELETIER, A. L. M., COMTE DE ST FARGEAU.

SALVIN, O.   *see* GODMAN F. and SALVIN, O.

SANDERSON, A. R.   Series of butterflies (*Delias*) and mimetic moths migrating together over a pass at Bukit Kutu, Selangor, Federated Malay States (1915–22).

SARAWAK MUSEUM, KUCHING   Orthoptera from various localities in Sarawak (1906–8).

SAUNDERS, E.   16 Buprestidae (by exchange, 1866).
British Hymenoptera Aculeata from his own collection (1897).
Rare British Hymenoptera (1899).
Collection of British Hemiptera (presented by G. B. Longstaff and E. B. Poulton 1910).
Hymenoptera *via* G. A. J. Rothney (1915–22).
Numerous other specimens over many years.

SAUNDERS, SIR S. S.   A few Albanian insects, mainly Hymenoptera, to J. O. Westwood (August 1849).

June 1884. Purchased from the Representatives of the late Sir Sidney Smith Saunders his entire Collection of English & Foreign Strepsiptera, occupying 2 large drawers—Price 25£. Also his Collection of briars, inhab^d by Insects from Albania and elsewhere classified in one of Standish's large boxes—also a quantity of split briars not arranged, in a glass covered case— Also, the Collection of the perfect insects, inhabiting the briars, pinned and classified (with their parasites) filling another of Standish's large boxes—& in a large glazed drawer. Also the Collection of Strepsipterous & other insects with their details mounted as 116 microscopical slides arranged in 7 boxes—& another Collection of specimens of these insects & of larvae &c. of Meloidae in 177 small bottles in spirits—Also a Collection of larvae & illust.[ns] of transformations of Meloidae &c. And three thick 4[to] Volumes filled with MSS notes of Observations on the above ment.[d] Collections.

Collection of Greek Hymenoptera (E. Saunders, January 1886, £30). Kept separate.

'Blind Cave Insects and other Articulata from the Caves of Adelsberg', in spirit (J. O. Westwood, n.d.).

SAUNDERS, W. W.   A few Hymenoptera, Hemiptera, and Neuroptera from Brazil to J. O. Westwood (December 1848).

Lepidoptera, chiefly diurnal, from tropical West Africa and Celebes, including *Papilio Neptunus* (1859).

Collection of Buprestidae mainly from Chile, and a few from Argentina and Brazil, Australia, Haiti, Corfu, New Holland, Congo, Cuba, and Fiji (1866).

Polybothris zygaena, Buprestis helopioides, and *Iridotania igniceps, moluccana, instabilis,* and *jansoni* of which the latter three species were collected by A. Russel Wallace at Kaisa, Ternate, and New Guinea (by exchange, 1866).

Symphaedra æropa and *Pieris* (1866).

Three Diptera from south Europe, 80 Diptera, and 10 Fulgoridae from general collection, 13 Diptera (May–November 1867, through F. Walker, at 10*d.* each, total cost, including some A. Russel Wallace specimens, £9. 5*s.* 10*d.*).

84 Homoptera and four Diptera (through F. Walker, June 1868, £4).

138 Coleoptera from New Holland, 38 Homoptera, 187 Diptera, two Hymenoptera (through F. Walker, November 1868, £2. 5*s.* 10*d.*).

15 Coleoptera from Brecknock Harbour, tropical Western Australia (January 1869, 18*s.*).

119 various specimens (from Janson out of Saunders Collection, June 1873, £1): 'Each marked with a lozenge shaped bit of red paper'.

Purchased by Mrs. Hope [1873] for the sum of £600 & presented to the Entomological Collection, Mr. W. W. Saunders' Collections of Orthoptera, Lepidoptera Heterocera & Illustrations of Economic Entomology.

Orthoptera Forficulidae 181 specimens, Blattidae 467, Achetidae 158, Locustidae 1756. European Collect.[n] 428. Chili & W. Austr. 10. Phasmidae 559, Mantidae 1104. Total 4663 contained in 59 large double Boxes.

Illustrations—885 specimens exclusive of specimens in nests, galls &c.

Lepidoptera Heterocera

Sphingidae	378 species	897 specimens	Contained in 156 large
Bombycites	3002 —	6423 —	double Standish Boxes
Noctuites	2368 —	4392 —	& 4 teak cabinets
Geomitrites	2100 —	3486 —	containing 28 long
Pyralites	1301 —	2272 —	corked & glazed
	9149 species	17470 specimens	drawers.

The Orthoptera contain Types of Bates, and Lepidoptera Heterocera those of F. Walker.

'A Box of exotic Lepidoptera Heterocera part of Mr. Saunders' Collection (sold at Mr. S's private sale) at Reigate' (Mr Janson 1874, £1, also at same sale some Coleoptera, 18*s.*). The Heterocera contained 20 specimens from Wallace Collection.

Exotic Coleoptera which includes a few Wallace specimens (September 1874, £3).

22 May [1875], Purchased from Mr. W. W. Saunders. His entire Collection of Exotic Hymenoptera price £150.0.0, arranged & named by Fred[k] Smith in 40 large double Standish Boxes—

Tenthredinidae	685	indiv[ds]	Siricidae	53		Evaniidae	136
Ichneumonidae	1537	—	Chrysid[ae]	259		Chalcid[ae]	212
Cynipidae —	15	—	Proctotrup[ae]	14	—	Formicidae	1958
Dorylidae —	38	—	Thynnidae	299	—	Mutillidae	459
Scoliidae —	558	—	Pompilidae	756	—	Sphegidae	546
Larridae —	235	—	Bembecidae	150	—	Nyssonidae	142
Crabronidae —	186	—	Cerceridae	152	—	Miscell.	60
Vespidae —	1298	—	Apidae	2667		Total	12415

at $2\frac{3}{4}$d each exclusive of boxes & charge of arrangement by F. Smith & description of n.sp. in Proc. Linn. Society.

Also 4 boxes of unarranged Hymenoptera.

In larger green Box 408 indiv[s] including 228 Apidae, 49 Formicidae, 44 Fossores, 29 Thynnidae, 19 Ichneumonidae &c.

In a smaller green box 126 individuals including 60 Ichneumonidae, 11 Proctotrupidae, 28 Chalcididae, 10 Stephanus, 7 Evaniidae &c.

In deal Box No. 195 325 individuals including 283 Formicidae, 13 Fossores, 13 Ichneumonidae &c.

In small Box of 'Minute Hymenoptera' 269 individ[s] including 239 Ichneumonidae, 17 ants, 9 Chalcid[d] &c.

Heteromera *via* H. E. Cox Collection (1915, 1922).

Arachnida and Scorpionidae *via* O. Picard (according to Horn, W. and Kahle, I., 1935–7. *Ent. Beih. Berl.-Dahlem*, **2–4**, i–vi, 1–536).

SAUSSURE, H. DE   Series of Orthoptera from various localities (1894).

SAVAGE, T. S.   Insects sent while he was a missionary stationed at Cape Palmas, West Africa, to F. W. Hope (1842–8), also a few to J. O. Westwood.

SAVIOZ, E.   Two boxes of Swiss Coleoptera (presented by H. W. Acland, November 1857).

SAVORY, T. H.   Collection of British spiders (1939). Remnants of his collection of spiders from Christmas Island (1941).

SAY, T.   Insects to Hope (*c.*1828).

The collections also contain a few specimens of Chalcidoidea from North America.

SCHÄFFER, G. A. W. HERRICH-   *see* HERRICH-SCHÄFFER, G. A. W.

SCHAUM, H. M.   North American insects, chiefly Coleoptera, sent to J. O. Westwood (1848).

Insects from Ceylon, Carniola, and Pyrenees (by exchange, 1862).

Two Coleoptera and one Arachnid (presented by J. O Westwood, 1863).

SCHAUS, W.   Series of named Geometrid moths from Tropical America (1901, 1903).

Two specimens of a dark-hind-winged species of *Protogonius* from Cayenne, characteristic of that part of South America (1906).

47 butterflies from Omai, British Guiana (1908).

170 butterflies and moths from neotropical localities (1913).

3169 specimens of Costa Rican Lepidoptera determined by donor (1925).

SCHEIBEL, T. K.   Coleoptera and a few other insects from Papua, New Guinea (*c.*1923).

SCHIÖDTE, J. H. C.   Insects to Westwood (1854).

SCHMELTZ, J. D. E.   Lepidoptera from Pelew (January 1864, 12*s.* 6*d.*).

Insects of various orders and one *Papilio schmeltzii* (July 1869, thalers 6.15 = 19*s*. 6*d*.).

SCHMIDT, H.   'Cecidiologisches Herbar', a small collection of plant galls transferred from Department of Forest Entomology, Oxford (August 1943).

SCHMIEDEKNECHT, O.   *see* MORICE, REVD F. D.

SCHÖNHERR, C. J.   *see* BADEN, F. and SOMMER, M. C.

SCHÖNLAND, S.   Small collection of the black ant *Paltothyreus tarsatus* captured at Mapellapveda, South Africa (1904).

SCUDDER, G. G. E.   Hemiptera–Heteroptera and other insects from Britain and Jersey, also insects of many orders from Tregaron Bog, Cardiganshire (1956, 1957).
Series of Hymenopterous parasites (Aphidiidae) bred from the Bedstraw Aphid from Wytham, Berkshire (1958).
139 British species of *Saldula*, and a collection of British Hemiptera–Heteroptera comprising some 3500 specimens representing half the British species (1960).

SEITZ, A.   Diptera from Algeria and Brazil, and mimetic insects from Brazil, Hong Kong, Singapore, and Ceylon (1913).

SELLON, H. S.   Collection of British Lepidoptera (presented by his mother and sisters, 1909).

SELOUS, F. C.   Insects collected in Matabeleland (1906).

SELYS-LONGCHAMPS, BARON M. E. DE   *see* DALE, J. C. and C. W.

SEMPER, C.   Butterflies from Luzon, Pelew Islands, and Mindanao (through Hewitson, 1864, 12*s*. 6*d*.):
Coleoptera collected by Semper (purchased from Higgins, 1866).

SENDALL, MRS I. W.   Two species of gregarious wasps and their nests and nests of *Oiketicus* sp., with living larva from Grenada, West Indies (1888).

SENDALL, SIR W.   Wasp nests, mud-nest of solitary bee, and seven Arachnids from Grenada, West Indies (1889).

SERVILLE, J. G. AUDINET   A few Hemiptera types and Orthoptera Phasmididae in Marchal and Van der Poll collections. Exotic bees *via* J. O. Westwood (1857).

SETH-SMITH, D. W.   Over 500 butterflies from the Gold Coast (1939–40).

SEVASTOPULO, D. G.   112 Asilidae with prey, from Darjeeling, India (1948).

SHARP, W. E.   Set of British and Irish Coleoptera Staphylinidae (1902, 1906). All named by donor.
A few beetles from Asia Minor and several European localities and two Carabidae from the Falkland Islands (1906).

SHAW, A. E.   Lepidoptera from Healesville, Victoria, Australia (1913).
Insects of various groups from Healesville (1914).
21 Blattidae (*Supella sepellectilium* Serv.) from house at Wynnum, Queensland (1939).

SHELFORD, R.   Vast collection of Bornean insects of many orders presented when he was Curator of Sarawak Museum (1899–1903). Includes the specimens illustrating his memoir *Mimetic insects and spiders from Borneo and Singapore* (*Proc. Zool. Soc. Lond.* **1902**, 230–84) and 10 Types or syntypes of new species.
Donated many other insects, especially Orthoptera Blattidae from collections sent to him for determination whilst working in the department.

SHELLEY, G. E.   Lepidoptera remaining in cabinets purchased from Stevens (June 1922, lots 125–31).

SHEPHERD, E.   22 Coleoptera by exchange (1858).

SHIPTON, S.   252 insects, almost entirely Lepidoptera, from Tucuman Province, Argentina (1904).

SICHEL, J.   98 specimens of French *Sphecodes via* J. O. Westwood (1866).

SIDGWICK, A.   British Lepidoptera (1897).
   3756 British Lepidoptera (presented by Mrs Sidgwick, 1921).

SIDGWICK, N. V.   British Lepidoptera (1929).

SIEBERS, H. C.   *see* HANITSCH, K. R.

SILBERMANN, G. H. R.   Insects to Hope (1824).

SIMMONDS, H. W.   Insects, chiefly butterflies, from Fijian group of islands (1919–35).
   Butterflies from New Zealand (August 1926).
   174 butterflies from Mauritius (1941).

SIMMONS, W. C. and MRS SIMMONS   Series of butterflies and a few moths from Kome Island, north-west Victoria Nyanza (1927).

SIMON, E.   *see* PICKARD-CAMBRIDGE, REVD O. (1917).

SIMON, M.   27 butterflies mainly from Oriental Region, new to, or poorly represented in the collection (1981).

SIMPSON, I. H.   Box of butterflies (1871).

SIMPSON, J. F. HOPE-   *see* HOPE-SIMPSON, J. F.

SIMPSON, J. J.   *see* COLONIAL OFFICE.

SJÖSTEDT, Y.   Duplicates of Blattidae from Kilimanjaro *via* R. Shelford, including syntypes (1906, 1907).
   See AURIVILLIUS, C. *and also* HANITSCH, K. R.

SKEAT EXPEDITION   *see* EVANS, R.

SLADEN, PERCY SLADEN MEMORIAL FUND   *see* BAKER, J. R. (1923, 1927).

SLATER, CAPT. M.   Indian Lepidoptera *via* J. O. Westwood (1857).

SLATTERY, H. F.   89 Peruvian butterflies, including five *Tithorea pavonii* and four *Heliconius charithonia peruviana* captured flying together in mimetic association, and a few other insects (1929).

SMEDLEY, N.   *see* HANITSCH, K. R.

SMITH, D. W. SETH-   *see* SETH-SMITH, D. W.

SMITH, F.   Several specimens identified as *Anthophorabia retusa* (Hymenoptera: Chalcidoidea) sent to Westwood, 1854. This was described in 1849 by G. Newport, whose original specimens were lost. Smith's material may be eligible for selection of a Neotype of Newport's species.

   1879, Oct. Purchased from W. Janson 993 Hymenopterous insects selected from the Collection of the late F. Smith as under

285 type specimens described & named by F. Smith at 1/-	14.5.0
708 other specimens not descr.[d] or named by F. S. at 6[d]	17.14.0
Carriage of Boxes containing the Collection & return	1.13.4
	£33.12.4

British Aculeates, including numerous Types, through R. C. L. Perkins (1930). Kept separate.

See also ROTHNEY, G. A. J.

SMITH, F. W.   British Lepidoptera selected from donor's collection (1949).

SMITH, H. GROSE   Insects from Madagascar, Solomon Islands, and Sierra Leone (June 1890, £16. 9s. 6d.).
Butterflies from Central China and Madagascar (1891, £1).
Two ♂ and two ♀ *Papilio merione* (1896, 10s.).
67 butterflies from the Tian Shan Mountains, Mongolia (1908, £3).

SMITH, H. H.   Butterflies from Central America in Godman–Salvin Collection (1896).

SMITH, H. R.   Insects from various British localities, a few Lepidoptera out of Swinhoe collection from Assam and Khasia Hills, and 633 insects of many orders from Russell, western Manitoba (1897).

SMITH, K. G. V.   *see* HAMMOND, H. E. (1956, 1957).

SMITH, W. C. LESTER-   *see* LESTER-SMITH, W. C.

SMITH, W. G. POGSON   British Lepidoptera (presented by Mrs Smith, 1915).

SMITHSONIAN INSTITUTION, US NATIONAL MUSEUM   Collection of birds' eggs, mainly collected by R. McFarlane, from Hudson Bay Co., via Smithsonian Institution (1870).
Marine invertebrates from New England coast collected by US Fisheries Commission, with more recent additions from inner edge of the Gulf Stream slope south of Martha's Vineyard collected in 1880 and 1881 (1884). The Crustacea mostly identified by Professor S. J. Smith.

SNELL, H. J.   *see* HANITSCH, K. R.

SNELLEN VAN VOLLENHOVEN, S. C.   *see* RIJKSMUSEUM VAN NATUURLIJKE HISTORIE, LEIDEN.

SOMEREN, R. A. L. VAN   Large collection of butterflies including Types and paratypes (1926, £25).

SOMEREN, V. G. L. VAN   Lycaenid butterflies attacked by lizards (1922).
12 families of *Papilio dardanus* each with its female parent taken at Nairobi, bred 1922 (1923).
Nymphaline butterflies from Jinja, Uganda, material of bionomic interest from Nairobi district, series of parasitic Hymenoptera bred from Lepidopterous larvae from Nairobi and Jinja district, and 15 butterflies from Marsabit, south-east of Lake Rudolph (1925).
Insects, mainly Lepidoptera, from Kenya Colony and Uganda, including some Types and paratypes (1926).
Charaxes etheocles from Kitale, Kenya Colony (1934).
Many other small donations.

SOMMER, M. C.   Box of insects from Island of Johanna to F. W. Hope (October 1838).
See also BADEN, F. and SOMMER, M. C.

SOUTHEY, MRS R. J.   91 butterflies, mostly Lycaenidae, from the Transvaal (1975).

SOUTHWOOD, MRS P.   Representative series of butterflies from the Congo Valley (1950).

SOWERBY, G. B.   Cabinet *via* Hope Deed of Gift (1849).
See also BAIRSTOW, S. D. (1891).

SPEYER, E. R.   Series of Odonata and Lepidoptera from Switzerland, Italy, Algeria, and Newfoundland (1911).

Spicer, Revd W. W.   Insects from Tasmania (*c*.1879).

Spiller, A. J.   Caffrarian butterfly *Hesperia canopus*, male and female (1891).

Spilsbury, Revd F. M.   British Lepidoptera (1878).
Endowment through Trinity College, Oxford, for upkeep of collection.

Spinola, Marchese M.   Two Pompilid wasps (Types) from his collection and labelled by Westwood as having come from Spinola.
Large series of bees sent to Westwood.

Stål, C.   Insects from Royal Museum, Stockholm (n.d.).

Standen, R. S.   Collection of butterflies from various localities including Thursday Island, Torres Strait, Fiji, Samoa, etc. (1906).

Standfuss, M.   Eight hybrid Saturniid moths (presented by F. A. Dixey, 1897).

Staudinger, O. and Bang-Haas, A.   94 Blattidae from various localities, including 20 new to science (1908).

Stephenson, W.   Box of insects and two Maori mats, one of which belonged to Chief of Tiaropa 'a monster who has frequently feasted on human flesh', and the other to Chief Etako, to Hope (1844).

Steven, C. von   Insects offered to Hope (1829).

Stevens, J. C. and S. (Stevens's Auction Rooms)   Exotic insects purchased for £12. 10s. 6d. and presented by F. W. Hope, including specimens collected by Fortune (China), Baly (Amazons), Wallace (Sarawak, Aru Islands) (June 1858).
Insects from Australia, Port Natal, Celebes (Wallace) (purchased and presented by Hope, November 1858).
Insects, mainly from the oriental region, some collected by Wallace (purchased and presented by Hope, August 1859).
Insects from New Holland and Borneo, including some collected by Wallace (purchased and presented by Hope, 1860).
Various insects, including a few collected by Wallace at Mysol and Waigiou (1862, £18. 7s. 9d.).
Various insects including some collected by Wallace (New Guinea, Mysol, Sumatra), Angas (South Australia), Damell (north and south Australia), and Fortune (*Papilio* from Japan) (1863, £24. 10s. 3d.).
Complete series of cocoons of silk-producing moths from all parts of the world, some moths and samples of silk, in five glazed cases, lot 165a, £2. 7s. 6d., also insects from Australia, lot 152, 5s., and two specimens of *Ornithoptera victoriae* ♀, lot 160, £1. 15s. (1888).
Lepidoptera from Tenerife, Andaman Islands, Queensland, Jamaica, China, Japan, Colombia, Bogotà, Honduras, Venezuela, Colorado, Malacca, Borneo, New Britain, Kashmir, etc. (1900).
75 specimens of the northern variety *hethlandica* of the 'Ghost swift', *Hepialus humuli* from Unst, Shetland (1900).
237 moths from Costa Rica, 391 Lepidoptera from southern India, and 94 Lepidoptera from Paraguay (1904, £2. 6s.).
Various unspecified insects were also purchased between 1863 and 1916.

Stevenson, W. S.   Box of Coleoptera to Hope (1835).

Stewart, J. H.   Box of insects sent from Calcutta to Hope (1832).

Stockley, Col. C. H.   101 butterflies from Mount Kenya and 41 from the Tana River, Kenya (1948, 1949).
49 butterflies from Kenya and Uganda (1952).

STONE, S.   Three bottles of larvae of *Vespa crabro* (1861).
Collection of leaves mined by larvae of Micro-Lepidoptera, Diptera, etc., hornet's nest, and a few specimens of Odonata (1862).
Specimen of *Acherontia atropos* with only one antenna and pupa case from which it was reared (1865).
Large series of nests of wasps and insects in spirit (presented by his relatives, 1866).

STONEHAM, H. F.   Miscellaneous insects from Uganda (1925).

STRACHAN, H.   212 insects of many orders and three Arachnida from Lagos, chiefly the Ogun River Basin (1900).

STRACHAN, S. L.   Boxes of insects, mainly Coleoptera, from Sierra Leone (1839). Some of the specimens were sent preserved in gin.

STRONG, W.   Insects in gum animé and amber to Hope (*c.*1837).

STROYAN, H. L. G.   120 microscope slides of British Aphididae (1960).

STURM, J.   Insects to Hope, mainly Coleoptera (1830–1).

SWINHOE, C.   Many donations of oriental Lepidoptera whilst working on the Collections, including 598 specimens of Pyralidae for the general collection of Lepidoptera Heterocera (1890–1902).
Some specimens through W. Holland, including 205 specimens of *Limnas chrysippus* and some allied forms from Afghanistan, Baluchistan, and India (1899).
See also SMITH, H. R.

SWINHOE, E.   Six specimens of African butterflies (purchased 1895).

SWYNNERTON, C. F. M.   Large collection of Lepidoptera from neighbourhood of and within the forest on Mount Chirinda, Melsetter, Gazaland, south-east Rhodesia, also some bionomic material (1910, 1911).
Insects, especially Lepidoptera, from Tanganyika Territory and south-east Rhodesia (1915–22).

SWYNNERTON, G. H.   194 insects of various orders from Cara Island, Argyllshire (1935).

SYKES, M. L.   Series of butterflies from West Africa, India, Burma, and Central and South America (1899).

TAYLOR, C. B.   202 insects of many orders from Jamaica (1903).

TAYLOR, E.   Collection of British woodlice, including material given to the donor by Dr A. Randall Jackson (1933).

TAYLOR, W. W.   Series of Cynipidae and Braconidae bred from oak galls from Wytham, and Braconidae bred from *Geometra papilionaria* from Tubney (1926).

TEMPLETON, R.   Insects, especially Coleoptera, from India, Ceylon, and Java *via* J. O. Westwood (1857).
Female white ants (pupae and winged), larva of *Dytiscus*, mole cricket, *Trombidium tinctorium*, common black house ant, in spirit, collected in Ceylon (presented by J. O. Westwood, 1863).
A few other insects from Mauritius.

TERRY, F. W.   British insects, also a few from Switzerland and India, including a large number of microscopic preparations (presented by his uncle E. Wray, 1931).

THAYER, A. H.   Butterflies from the eastern United States (1899).
Many insects from eastern United States (1902).

THOMAS, R. H.   Collection of Hymenoptera and their mimics from Salisbury, Mashonaland (1900).
   Insects from Colombia (1925).
   35 Coleoptera and 11 Rhopalocera from River Tigre area, South America (1936).
   142 Lepidoptera from Ecuador including many rarities and 37 other insects (1940).

THOMASSET, H. P.   *see* HANITSCH, K. R.

THOMSON, A. W.   Iceland collection of Agnes W. Thomson, 1911 (n.d.).

THOMSON, J.   Specimens collected in the River Gaboon area, Africa (September, 1857).

THORNLEY, REVD A.   Small collection of Coleoptera from various localities world-wide (1903).
   See also IMRAY, MRS R.

THORELL, T.   *see* PICKARD-CAMBRIDGE, REVD O. (1917).

THOYTS, MISS E. E.   Aculeate Hymenoptera and other insects from Berkshire and Yorkshire (1897, 1898).

THWAITES, G. H. K.   Specimens of the elephant fly of Ceylon (*Pangonia*) (presented by J. O. Westwood, September 1857).
   Insects of various orders collected by Mr Thwaites, Curator of the Botanic Garden, Colombo (presented by J. O. Westwood, 1861) (Arachnida *sens. lat.*, in spirit).
   'Various objects' and bottle of Crustacea from Ceylon (through A. Günther, 1871).

TILLYARD, R. J.   Primitive New Zealand moths (1915–22).

TODD, V. E.   12 jars of named Harvestmen (Phalangids) from Wytham Wood, Berkshire (1948).

TOMLINSON, W.   Twigs and leaves of coffee-tree infested with *Coccus coffeae* (1889).

TRIMEN, R.   Set of the butterfly *Melanargia occitanica* from Hyères, south-east France (1899).

TRISTRAM, CANON H. B.   291 insects of many orders from Palestine captured by the donor in 1863–4 and by O. Pickard-Cambridge in 1865, and borrowed by J. O. Westwood for study (formally presented to Hope Department 'in which they have rested so long', 1897).

TROUGHT, T.   Bred series of parasitic Hymenoptera and Diptera mainly from Tysoe, Warwickshire (1949).
   Six *Thaumetopoea jordana* from Jordan Valley (1953).

TUBBS, REVD F. D.   Small set of Lepidoptera from Mercedes, Buenos Aires, Argentina (1897).

TURLEY, R. T. and MRS TURLEY   194 insects of many orders, including an Arachnid and several Myriapoda, from Manchuria (1898).

TURNER, MISS E. M.   Representative collection of British dragonflies and earwigs (1938).

TURNER, G.   Mutillidae captured at Port Mackay, Queensland, in 1900 (determined and presented by his brother R. E. Turner, 1907).

TURNER, H. J.   Collection of Lepidoptera from various parts of the world (1949, 1950).

TURNER, J. A.   Insects to Hope (1829), including some South American Coleoptera.

TURNER, LIEUT.   25 butterflies and five moths from Zanzibar (1884).

Tuxen, S. L.   Six Coleoptera, four Lepidoptera, and eight Diptera from Greenland (1939).

Tylden, Revd W.   Australian and Japanese insects (1872).
Collection of Coleoptera, mainly Curculionidae (presented by Mrs Tylden, 1875) (Plate 17).

Tylecote, E. F. S.   Collection of British and European Lepidoptera, also specimens from Texas, Jamaica, Ceylon, and India (1933).

Tytler, Maj.-Gen. Sir H. C.   1047 butterflies, mainly from the East (purchased 1941). Includes some paratypes and many rarities.

Vallentin, R. and Mrs Vallentin   Collection of insects from the Falkland Islands (1922). Kept separate.

Van der Poll, J. R. H. Neerwort   Collection of Orthoptera and Chevrolat's collection of Coleoptera Carabidae (presented by E. B. Poulton, 1909, who had purchased them for £40).
6000 butterflies, chiefly Malayan (P. I. Lathy, 1910, £30).
Malayan Hymenoptera, mainly Javanese (presented by G. A. J. Rothney, 1911).

Varley, G. C.   Two slides (four specimens) of the Aphid *Hysteroneura setariae* from Magnetic Island, Queensland (1973).
Many British specimens, also insects from Ceylon, Bahrain, northern Finland, and Malaysia during his tenure of the Hope Chair.

Venables, L. S. V.   *see* Lack, D. and Venables, L. S. V.

Verkrüzen, T. A.   Collection of Coleoptera £18 and miscellaneous insects 7s. 6d. (purchased July and August 1888).

Verrall, G. H.   Bigot collection of Asilidae, including some from his own collection, was presented by Verrall in 1908.
His main collection of Diptera passed to J. E. Collin.

Verrall, G. H. and Collin, J. E.   Collection of Diptera, comprising some 140 000 specimens of British, Palaearctic (including Bigot and Kowarz collections) and some exotic species. Contains many Types of species described by Collin, Verrall, Bigot, Macquart, Edwards, Melander, and others. Presented by J. E. Collin on condition that it shall be known as the Verrall–Collin Collection and kept intact for 50 years (1967).

Main British Collection
Includes Types, generally as syntypic series, some being Types of fungus gnats, described by F. W. Edwards and the best material from A. E. J. Carter collection.

Main Palaearctic Collection
An amalgamation of Bigot Palaearctic collection and Kowarz European collection purchased by Verrall, and other material acquired by him and Collin.

Exotic Collection
Contains Bigot Types and some of Macquart.

Duplicate Collection
Contains Verrall, Collin, and Carter specimens; also miscellaneous named and unnamed material.

Vevers, H. G. and Evans, F. C.   87 Coleoptera and Diptera from the island of Myggenaes in the Faeroes (1938) (dried and spirit material).

Viktorov, G.   17 Ichneumonidae bred from known hosts in the USSR (1971).

Villa, A. and G. B.   Coleoptera to Hope (May 1844).

Villiers, F. de   Insects to Hope (December 1836).

Vinall, Miss G.   Lepidoptera from the Belgian Congo (1928–35).

Wager, L. R.   217 insects of all orders from British Expedition to East Greenland 1935–6 (1936) (includes four jars of spirit material).

Waghorn, H.   Zambezi insects of various orders (purchased 1864, £1. 10s.).
Coleoptera collected at the Zambezi (1864).

Wailes, G.   28 male *Anisolabis maritima* taken in profusion on the Northumberland coast (September 1857).

Wain, Father F. L.   2200 bees from India and the Himalayas (1979).

Wainwright, C. J.   47 British Tachinid flies (1928).

Walcot, Miss F.   Specimens of cherry and plum-tree shoots from Canada injured by gall-forming Aphid (1890).

Walker, C. R.   Lepidoptera from northern Nigerian localities (1915–22).

Walker, F.   84 microscope slides containing a series of British Aphids mounted and named by him (presented by the Entomological Society of London, July 1863).
Specimens from his collection (purchased at auction June 1877, £4. 1s.).
Collection of Aphides (F. A. Walker, 1879, £5).
Hymenoptera *via* G. A. J. Rothney (1915–22).
The Collections also contain over 2000 Walker species of moths including Types *via* W. W. Saunders (1873), 70–80 Hemiptera–Homoptera Types, and one box of Walker and Meade British Diptera (some may be Types).
Some syntypes of Hymenoptera Chalcidoidea taken by him at Alten, Finmark, Norway; and specimens of Plecoptera, Ephemeroptera, Neuroptera, and Trichoptera collected by him during his journey from Alten to Stockholm, in 1836, in Dale collection (see Graham, M. W. R. de V., 1979, *Entomologist's Gazette*, **30,** 18).
See also Kollar, V.

Walker, Revd F. A.   Iceland collection 1889 (presented by G. B. Buckton, 1907).

Walker, J. J.   Lepidoptera, Coleoptera, and a few oceanic Hemiptera captured during voyage of HMS *Penguin*, 1890–3, chiefly oriental and Australian (1896).
Series of *Hesperia lineola* from Sheerness (1896).
Eight specimens of two species of Forficulidae new to Britain and rare Coleopterous insect from Chatham (1897).
Insects from Greece, Malaya, northern Australia, and south-east China, also Lepidoptera and Coleoptera from various British localities (1899).
Insects, chiefly Lepidoptera, from Cape Verde Islands, New Zealand, Australia, New Hebrides, New Caledonia, Torres Islands, Loyalty Islands, Tahiti, New South Wales, Chile, Panama, etc. (1904).
Insects, chiefly butterflies, from localities world-wide (1905).
Moths from various American localities, also some Australian Hesperiidae (1910).
Insects from world-wide localities (1913).
Collection of New Zealand Coleoptera (1915–22).
Butterflies captured during voyage of the *Ringarooma* in 1900, from New Caledonia, Banks and Torres Islands, New Hebrides and Loyalty Islands, etc. (1921–3).
34 marine Hemiptera of the genus *Halobates* from mid-Pacific, Marquesas, Nimrod Sound, China, New Hebrides, and Torres Islands, 1882–1900 (1925).
291 moths collected at Gibraltar and north coast of Africa, also 332 Hesperiidae collected at North American and South American ports (1933).
Butterflies, mainly Lycaenidae, from coast of South America (1935).

Collection of British Coleoptera and Lepidoptera, 87 boxes of foreign butterflies, and 42 boxes of Australian and New Zealand Coleoptera, books, and separata (1939).

Donated much British material while associated with the department. Statement by Hope Professor in Annual Report: 'The British Beetles will be kept intact, as he wished, as far as possible.'

WALKER, SIR PATRICK   Cabinet *via* Hope Deed of Gift (1849).

Cabinet containing English Lepidoptera, Diptera, and Hymenoptera *via* J. O. Westwood (1857).

WALLACE, A. RUSSEL   Hymenoptera, etc., from Pará, Brazil (purchased by J. O. Westwood, 1848).

70 specimens of British Lepidoptera (1858).

20 diurnal Lepidoptera from Celebes (November 1858), insects from Sarawak and the Aru Islands (June 1858), Borneo, Amboyna, Dorey, Batchian, Ternate, and Gilolo (August 1859), and from Borneo and Sarawak (June 1860). Purchased from Stevens by F. W. Hope.

Squilla desmarestii from Isle of Wight (1861).

Insects from Mysol and Waigiou (purchased from Stevens, 1862).

Two larvae and one pupa of *Mormolyce phyllodes* (presented by J. O. Westwood, 1863).

Insects from Sumatra, New Guinea, and Mysol (purchased from S. Stevens, July 1863).

Entire private collection of Melolonthidae, Rutelidae, Trogidae, Aphodiidae, and genus *Valgus* (514 specimens) from the Malayan Archipelago, also his private collection of Eumorphidae (201 specimens), Pselaphidae, and Scydmaenidae (29 specimens), from the same islands (May 1865, £28. 16s.).

Private collection of Clavicorn Coleoptera made in the Malayan Archipelago (May 1866, £10).

Three specimens of *Iridotania* from Kaisa, Ternate, and New Guinea (through W. W. Saunders, 1866).

Entire private collection of Cleridae formed in the Malayan Archipelago containing 697 specimens, also his collection of Staphylinidae containing 523 specimens (1866, £35).

'Purchased from Mr. Walker (on acct of W. W. Saunders) 73 Diptera, 29 Homoptera from the Malayan Archipelago collected by A. R. Wallace & described by Messrs. Walker & Stål, from Mr. Saunders Collection' and '15 Cercopidae from ditto' (April 1867 at 10d. each).

Various insects from Wallace collection (Mr Higgins, 1868, £2. 7s., 7s., 12s., £1. 12s., 3s., 4s.).

Private collection of Heteromerous Coleoptera (December 1869, £32. 10s.).

271 butterflies (Mr Hewitson from collection of A. R. Wallace, May 1871, £26. 14s.).

Assorted insects selected from Wallace collection (Mr. Higgins, May 1871, £8. 19s. 9d.).

Malay collection: three *Sospita*, two *Mycalesis*, and 40 small butterflies chiefly *Polyommatus* (from Mr Hewitson, July 1874, £1. 4s.).

A few Coleoptera amongst specimens purchased from W. W. Saunders (1874).

'1876. April. His private Collection of the following families of Malayan Coleoptera. Price £40.0.0. Anthribidae 1080 specimens, Brenthidae 605, Malacodermata 909 + 35 half eaten individuals, Hydrophilidae &c. 107—Passalidae 86 in., Coprides 199, Dynastidae 19 Oryctes—68 other genera. Total 3073.'

104 Carabidae from Malay Archipelago, included in lot 345 purchased March 1877 at sale of Edwin Brown's collection.

Butterflies from Malay Archipelago in Godman–Salvin Collection (1896).

Rare Castniid moth, *Castnia therapon*, together with pupa case, caught in donor's Orchid-house at Broadstone, December 1908. The pupa case was found among roots of a *Stanhopea*. The orchid and undoubtedly the larva within it had originally come from Santos, Brazil (1909).

WALLACE, E. A.   27 butterflies from Colombia (1886, £1. 12s. 6d.).

WALLACE, W.   22 butterflies from Ocaña, Bogotà (November 1878, £1. 5s.).

WALLER, W. E.   Lepidoptera from Iraq and the Pyrenees collected c.1918 and 1920 sent to Poulton (n.d.).

WALLIS, H. M.   Two specimens of *Parnassius mnemosyne* from the Pyrenees (1895).
31 Lepidoptera from the Italian Alps (1897).
56 butterflies, three moths, and a Dipteron from Switzerland (1899).
Lepidoptera from Fionnay Valley, Switzerland (presented by H. M. and A. Wallis, 1900).
Lepidoptera from Algeria and Bernese Oberland (1910).

WALSINGHAM, LORD   Preserved larvae of several British moths (1889).
Six specimens of the rare moth *Eupoecilia degreyana* from Thetford (1900).

WALTER, J. M.   *see* PARKER, D. E.

WALTERS, O. H.   50 beetles and 19 fig insects from Pinetown, Natal (1965).

WALTON, J.   Several lots of insects including a large box of Staphylinidae (from sale of his collection, 1863, 10s. 6d.).

WARD, H.   12 fine Dynastidae and Lucanidae from Darjeeling (1899).

WATERFIELD, A.   119 specimens of Lepidoptera from Demerara, British Guiana (1897).

WATERHOUSE, G. A.   Australian Pieridae (1911).

WATERS, E. G. R.   Extensive series of Micro-Lepidoptera, many bred from larvae collected in Oxford district, six albums of leafmines, British and some European Macro-Lepidoptera, and some British dragonflies (1930, £100).

WATKINS and DONCASTER   Two *Charaxes protoclea* (1880, 10s.).
Pieridae to supply important deficiencies in collection (1893, £2. 4s.).
Lepidoptera (almost exclusively Rhopalocera) and one specimen of a Coleopterous species, chiefly from Venezuela and Colombia, which formed the nucleus of a special collection illustrating mimicry (1895, £21, £4. 11s.).
Additional specimens for mimicry collection (1897, £4. 17s. 5d. including some apparatus).
Pair of a rare Danaine butterfly from Madagascar (1898).
Lepidoptera from Peru, Upper Amazon, Ecuador, Brazil, Colombia, Borneo, West Africa, and Madagascar (1901, £11. 19s., including some apparatus).
Butterflies for mimicry series (1902, £5. 2s.).
87 mimetic Lepidoptera from South American localities and 15 varieties of *Kallima paralekta* from Java (1903).
Butterflies (1889, £6. 8s.).
Various unspecified insects were also purchased between 1881 and 1890.

WATSON, CAPT. E. Y.   88 butterflies collected in Burma and 41 from Indian localities, mainly Mysore (presented by Trustees of British Museum (Natural History), 1901).

WATSON, [J.]   458 butterflies from Watson collection (Hewitson, May 1871, £19. 4s. 6d.).

WATSON, J. H.   Box of Saturniid cocoons (1910).
   Saturniid moths with their cocoons and four *Hypolimnas bolina* from Fanning Islands (1915–22).

WEALE, J. P. M.   Insects of all orders from Caffraria (March 1878, £4).

WEIR, J. JENNER   Series of Danainae and Acraeinae from various localities, from his collection (purchased at Stevens, June 1894).

WELLMAN, [F. C.]   Miscellaneous insects from Angola (n.d.).

WELLS, H. G.   Entire collection of exotic and British insects (presented by his son A. G. Wells, July 1859).

WESLEY, J. S.   27 English Lepidoptera (1858).

WESMAEL, C.   A series of 47 named Braconidae. Below the series is a label in Westwood's handwriting 'Types of Ichn^{ae} adsciti. from Wesmael. Belgium' (n.d.).

WESTERMANN, B. W.   Insects to Hope by exchange (1827–44).
   Insects of all orders from world-wide localities (1858).

WESTWOOD, J. O.   Westwood's entire insect collection was purchased by F. W. Hope and presented 31 July 1857. Part of it had been acquired from other famous collectors who are also listed under their respective names. Westwood was assiduous in purchasing even small collections which he recognized as being important. The following is the list as entered in his journal:

A Walnut wood Cabinet (originally Mr. Haworth's) containing 33 drawers filled with exotic insects of various orders.
A Buhl inlaid Cabinet of 26 small drawers containing English Hymenoptera Hemiptera &c.
A Mahogany Cabinet of 30 small drawers (originally Sir Patrick Walker's) containing English Lepidoptera Diptera Hymenoptera & Illustrations.
A Cabinet of twenty drawers made of Indian Wood (Sir J. B. Hearsey's) containing Indian Lepidoptera & Illustrations of English Insects Economy &c.
A Cabinet of 12 small drawers with talc tops (formerly Captain Boys') containing Indian Hymenoptera Diptera Hemiptera &c.
A Cabinet of Eight Drawers made of very heavy Indian wood (Sir J. B. Hearsey's) containing Indian Lepidoptera.
A Cabinet of Twelve Drawers (deal, with 2 drawers beneath for bottles in compartments) containing Indian insects chiefly Lepidoptera.
A set of fifteen mahogany drawers containing Exotic Hymenoptera Orthoptera &c.
A set of 9 large deal stained drawers containing Exotic Coleoptera.
A set of 13 deal store boxes covered with green cloth containing British Coleoptera.
A set of five double boxes containing a fine series of Mexican insects of all orders collected by Mr. Coffin.
Two large double boxes with Exotic Bees from Collections of Latreille & S! Fargeau & Serville— named typically.
One box of insects of various orders from Adelaide Australia sent by Mr. Wilson.
Four boxes of insects of Coleoptera & Hymenoptera selected from Mr. Bateman's Collection made at Melborne Australia.
Three Double boxes covered with green calico containing exotic Heteromera & Clavicorn Coleoptera.
A set of thirteen large double store boxes covered with green calico, containing exotic Coleoptera Orthoptera (& one, illustrations of economy).
Three store (double) boxes covered with salmon paper containing British Orthoptera & Exotic. Sphingidae & Nocturnal Lepidoptera.
Three large double boxes with exotic Diurnal Lepidoptera.
One very large double box with Indian Hemiptera (Dr. Bacon).
One box with Exotic Heteroptera (returned from Dallas).
One large double box painted green with Lucanidae & Rutelidae.

One large double box (mahogany) with about 800 species of N American insects (Col & Hem) named by Fitch Java Lepidᵃ frm Dr. Horsfield & large Nocturnal dᵒ Natal.
One large double store mahogany with Java Lepidoptera from Dr. Horsfield & Indian dᵒ (Captⁿ Mortimer Slater).
One large double store box Mahogany with Assam Lepidoptera Major Jenkins.

One	dᵒ	dᵒ	dᵒ	dᵒ with Cicadae & Fulgoridae.
One	dᵒ	dᵒ		with apterous insects.
One	dᵒ			Lepidoptera Cape Palmas.
One	dᵒ			Exotic Locustidae.
One	dᵒ			British Ichneumonidae named by Desvignes.
One	dᵒ			Achetidae & Gryllidae Exotic.
One	dᵒ			Butterflies Mussooree.
One	dᵒ			Indian Nocturnal Lepidoptera.
One	dᵒ			Ceylon Coleoptera Wollaston.
One	dᵒ			Illustrations of habits (mahogany box).
One	dᵒ			dᵒ (French Carton).
One	dᵒ double			Hymᵃ & Coleopt frm Turpentine Raddon.
One	dᵒ			Dynastidae.
One	dᵒ			Crustacea

Six boxes with dissections of insects by Latreille.
Seven deal Indian drawers with insects from India Java Coleoptera (Hearsey, & Templeton) Hymenoptera (Boys) &c. also Heliconiidae
Ten Carton boxes Indian Lepidoptera Hemipt. Orthopt & Hymenᵃ (Hearsey).
A large number of packages containing illustrations of Economy.
Several hundreds of bottles with insects, larvae &c in spirits.

Specimens of the Australian Manna and the insect by which it is produced (*Eurymela fenestrata*) (1865).

Bipalium kewense (1889).

A large amount of material was presented to Westwood during his tenure of the Hope Chair and this was usually given to the Collections.

WHEELER, REVD G.   246 butterflies collected 1901–14, nearly all from Swiss localities (1923).

WHITE, C. M. N.   36 butterflies from Crete (1937).

90 butterflies mainly *Acraea* species, from Northern Rhodesia (1948) and 33 from north-west Rhodesia (1949).

WHITE, MRS HOLT   Insects (chiefly Lepidoptera), four centipedes, and one spider from Tenerife (1895).

WHITE, W. WALMESLEY   Collection of butterflies and moths from Tenerife (1909). Kept separate, but some Lycaenids incorporated in general collection.

WHITEHILL, COL. S.   Insects to Hope from India, Cape of Good Hope, Pau, and the Pyrenees (1831–7).

WHITMARSH, REVD T.   Four trays of microscope slides of Hymenoptera Cynipoidea, and some Chalcidoidea, and three drawers of Cynipid galls (1881).

WIGGINS, C. A.   Collection of butterflies from Toro, Uganda, Lake Victoria Nyanza, and Mombasa, including a fine series collected by Major Rattray in Toro, of which some species are new to science (1902, 1903).

Butterflies, chiefly Pierinae, captured at Taveta, British East Africa (1905).

Lepidoptera and dragonflies from Kisumu on the north-east shore of Victoria Nyanza, insects of many orders from Mombasa, British East Africa, and other localities in Usoga on the north shore of the Victoria Nyanza, and a set of 153 male Pierines captured in a single sweep of the net at Jinja by the Ripon Falls, Victoria Nyanza (1906).

Mimetic butterflies and models from a tropical forest near Entebbe (1909, 1910, 1911).

Additions to collection of Uganda insects presented in earlier years, also large collections found abandoned during the First World War (1915–22, 1925).

WILEMAN, A. G.   Small series of butterflies from Formosa (1904).

WILLIAMS, C. B.   134 butterflies from Peru (1945).

Over 500 butterflies from various localities in tropical America and 11 from East Africa (1947).

Migration records and supporting voucher specimens (*c*.4500) presented by Mrs Williams (1977).

WILLIAMS, H. B.   72 specimens of *Pieris napi* bred from Donegal strain (1942).

WILSON, C. A. (cousin of A. Russel Wallace)   Insects of various orders from Adelaide, Australia *via* J. O. Westwood (1857).

Calosoma Curtisii Hope, larva and imago (1863).

WILSON, Miss D. R.   72 butterflies, 10 moths, and five dragonflies from São Paulo, Brazil (1913).

WILSON, J.   Insects from Java sent in 1834 and purchased by Hope (1835, £5. 17*s*. 6*d*.).

WILSON, Miss J. M.   10 specimens of Dermaptera from Nepal (1979).

WILSON, Capt. R. S.   Collection of insects, chiefly butterflies, from the Nuba Mountains, Province of the Sudan, also a fine series from the Western Desert Province of Egypt (1915–22). This material was collected *c*.1904, 1917–20, and includes specimens collected by Major C. C. Marshall and A. L. Kent-Lemon.

WIMBERLEY, H. I. A.   Insects from Southern Nigeria (*c*.1912).

WINCKLER, A.   Series of cave beetles from collection of A. Winkler (presented by C. Mackechnie Jarvis, 1963).

WINCKWORTH, Col. H. C.   5095 butterflies from South India and Ceylon (1938); a few of these were collected by his brother, R. Winckworth.

Transfers of scales from wings of butterflies from the Andaman Islands (1939) (in Archives).

87 butterflies from Ceylon and 37 from the Andaman Islands (1940).

134 butterflies from the Andaman Islands and two from India and Burma (1942).

124 butterflies from the Andaman Islands and India (1949).

WINTHEM, W. VON   Insects to Hope by exchange (*c*.1832–5).

WOLLASTON, A. F. R.   *see* PLAYNE, H. C.

WOLLASTON, Mrs E.   Six Micro-Lepidoptera captured by her on St Helena (1879).

WOLLASTON, T. V.   Ceylon Coleoptera *via* J. O. Westwood (1857).

Collection of land and freshwater shells of the Madeira Isles (purchased by subscription among the members of the University, 1861, £132).

Cabinet of 11 drawers containing collection of Madeiran insects, mainly Coleoptera, including many Types (purchased and presented by F. W. Hope, 1860, £300). Kept separate.

Additional material presented by Wollaston for Madeira Collection (1862).

Cabinet of 11 drawers containing Coleoptera of the Canary Islands (purchased and presented by Mrs Hope, 1863, £200). Kept separate.

Coleoptera from Madeira, Canaries, and Salvages (1865).

Box of Cape Verde Coleoptera (November 1867, £5).

Selection of Madeira Coleoptera, second collection (October 1870, £1. 2*s*. 6*d*.).

Cabinet and collection of microscopic preparations of insects (chiefly Madeiran) prepared by Wollaston for his various entomological works (Mrs E. Wollaston, March 1878, £20).

81 St Helena Coleoptera (Janson, December 1878, £3. 15*s*. 9*d*.).

30 Coleoptera taken on St Helena (presented by J. C. Dale, 1881).

In the Hope Library copy of Wollaston's *Catalogue of the Coleopterous insects of the Canaries in the Collection of the British Museum*, 1864, the second paragraph of the preface has been annotated by Westwood thus: 'The first selection of specimens from the Collection has been arranged by Mr. Wollaston and has been purchased by the Trustees of the British Museum, a second selection was purchased by Mrs. Hope for £200 and presented to the Oxford Museum. J.O.W.'

The Dale Collection contains four drawers of beetles from Madeira, Cape Verde, Canary Islands, and St Helena, also five drawers of shells from the Atlantic Islands (many from Porto Santo, some from the Desertas, Tenerife, Grand Canary, Gomera, and Lanzarote).

Wolley, J.   Birds' eggs collected by the late John Wolley, Jr (presented by F. W. Hope, June 1860).

Wood-Mason, J.   Butterflies from Assam and Port Blair (1882).

Wood, R. C.   37 butterflies from Nyasaland (1938).

72 butterflies, four moths, and one Nemoptera (Neuroptera) from Nyasaland and Cape Province (1939).

Eight butterflies from the Cape and Nyasaland (1940).

Lepidoptera damaged by birds or lizards (1941).

A few butterflies new to collection from South Africa and Nyasaland (1942).

Woodford, C. M.   Butterflies collected by him in Solomon Islands in Godman–Salvin collection (1896). The Collections also contain a few specimens captured by Woodford on Guadalcanar, purchased by J. O. Westwood.

Woodforde, F. C.   British Lepidoptera presented over a number of years while re-arranging the British collections, also many other insects (1910–26).

Insects, mainly Lepidoptera, from Passavant, eastern France (1922, 1923).

Woodforde, F. H.   118 glass cases of mounted birds, collected for the most part in Somersetshire and on Lundy Island between 1837 and 1867 (presented by F. C. Woodforde, 1911, with additions, 1912).

Conditions respecting a collection of British Birds deposited in Oxford Univ. Museum by F. C. Woodforde. The collection to remain unbroken, in their cases as now, until Jan. 1st 1921. If claimed by my son Geoffrey Cotton Woodforde, now of Gatooma, S. Rhodesia, S. Africa, before the expiration of that time, they are to be handed over to him on or before that date. If unclaimed by him they are to become then the absolute property of the Museum Delegates, provided that each specimen is distinctly labelled as having formed a part of the collection of my father, the late F. H. Woodforde, M.D. of Taunton. Francis Cardew Woodforde. Oxford. March 15th, 1911. [Countersigned by E. B. Poulton.]

The collection, officially presented in 1920, was later transferred to the Zoological Collections and subsequently donated by them to Bristol University Museum in 1951.

Woollatt, L. H.   About 12 000 specimens of British Hymenoptera, mainly sawflies and Aculeates, and Diptera, also blown Lepidopterous larvae from Finland prepared by E. O. Peltonen, and selected works from his library (1974).

Wordie, J. M.   Specimens from his expedition to north-west Greenland and east Baffinland, 1934. (See Longstaff, T. G., 1936. *Ann. Mag. nat. Hist.* **(10) 18,** 531–3.)

Wormald, P. C.   21 specimens of British Trichoptera (1863).

Worrall, Miss K.   Butterflies from Calcutta, Mandalay, and Upper Burma, some of bionomic interest (1904).

Wroughton, R. C.   Series of ants preserved in spirit, mainly from Oriental Region but includes some from Australia, South Africa, and America (1907). The material has been studied by Forel and Bingham and contains many syntypes.

Wykes, Miss U. M.   *see* Lupton, Miss P. M.

Yerbury, Col. J. W.   102 Lepidoptera from Hyères, South of France (1898).
   Diptera captured in Aden, which includes Syrphidae described by Verrall, and Asilidae by van der Wulp, also relevant manuscript notes (1899).
   Oestrus ovis from Simla and three adults and four puparia of *Cephalomyia maculata*, the rare gadfly of the camel, bred from larvae obtained in the neighbourhood of Aden (1899).
   British Diptera and examples of Diptera resembling Hymenoptera from Scotland for the mimicry series (1899).
   77 butterflies from southern Spain, a few Hymenoptera Aculeata and mimetic Diptera from County Kerry, Ireland, also models and mimics of same orders from Inverness, Cromarty, and Sutherland (1901).
   Insects of various orders from British localities, including additions to bionomic series from Herefordshire and neighbourhood of Barmouth, also a few Diptera from France, Spain, Portugal, and Aden (1902).
   Insects of various orders especially Diptera from British localities and a few from Arabia and Ceylon (1903).
   British Diptera (1904).
   130 Asilid flies from India and Ceylon and a series of British flies named by himself, Verrall, and Collin (1905).
   137 Coleoptera from Porthcawl, South Wales (1906).
   187 Diptera from Portugal, southern Spain, Hyères, and the Tyrol (1907).
   British Diptera, including many Dolichopodidae, determined partly by Verrall, Collin, and donor, also British insects of other orders including Hymenoptera Aculeata determined by E. Saunders (1907, 1909–10, 1911).
   Diptera, chiefly Bombyliidae, from Aden (1913).
   Books and offprints to library (1927).

Yonge, Revd J. U.   Lepidoptera from Madagascar and a few other insects (1906–14).

Young, Capt. G.   *see* Doncaster, L. (1927).

Yule, Col. J. B.   62 butterflies collected in Nyasaland (1900).

Zimmerman, K. C. A.   Insects from South Carolina and Pennsylvania to Hope (1835, price 'two Louis d'or' = £1. 3*s*. 4*d*.). Pennsylvania specimens 'marked with gold-paper'.

Zoological Museum, Tring   Five small Coleopterous groups exhibiting mimicry, or more probably warning colours, from French Congo, Loanda, Sierra Leone, Sumatra, and south-east Brazil, together with a similar association between Lycid beetles and their mimics from north-west Ecuador, and a series of butterflies from Abyssinia (1903).
   Two male specimens of *Papilio laglaizei* and two of its model, the Uraniid moth *Alcidis aurora*, from Dutch New Guinea, and *Papilio menestheus* from the Gold Coast, also 70 Longicorn beetles from Ethiopian Region including syntypes described by K. Jordan (1908).
   Also numerous other donations.

Zoological Society, Amsterdam   Insects from Japan, Java, Celebes, etc. (1858).

Appendix C.   Transcript of Mr Hope's Deed of Gift.
Oxford University Archives W.P. α/40/1

To all to whom these presents shall come The Reverend Frederick William Hope of Upper Seymour Street Portman Square in the County of Middlesex and late of Christ Church Oxford M.A. Sends Greeting. Whereas the said Frederick William Hope is the owner of a valuable Entomological Collection Library of Natural History Plates Engravings and other articles and effects now at N.° 37 Upper Seymour Street aforesaid which or the chief part thereof are particularized in the Schedule hereunder written and he is desirous of assigning and making over the same to the Chancellor Masters and Scholars of the University of Oxford for the purposes and upon the terms and conditions hereinafter mentioned. Now these Presents witness that in pursuance of and for effectuating and carrying out his said desire and out of the respect and regard he entertains for the said University the said Frederick William Hope hath given granted bargained assigned transferred and set over And by these presents Doth give grant bargain assign transfer and set over unto the Chancellor Masters and Scholars of the University of Oxford and their successors the Entomological Collection Library of Natural History Plates Engravings and other articles and effects hereinbefore mentioned and referred to And all the right title interest benefit property claim and demand whatsoever of him the said Frederick William Hope therein and thereto To have and to hold the same unto the said Chancellor Masters and Scholars and their successors absolutely and for ever But nevertheless upon and for the trusts and purposes and subject to the conditions following (that is to say) First, That a suitable Building or Rooms shall be provided within the University for the reception of the said Collection Library Articles and Effects as soon as conveniently may be arranged either at the expense of the said University or by means of Contributions from other sources. Second That the said Collection Library Articles and Effects shall be made practically useful as a means of extending and improving a knowledge of the Entomological Department of Natural History and that for that purpose the same shall at all seasonable times be free of access to Members of the University and other persons especially to learned Naturalists and other scientific persons of Foreign Countries subject however to such regulations as the Curators for the time being shall think fit and it being hereby declared to be the wish and desire of the said Frederick William Hope that so far as may be found practical and convenient the said Collection shall be opened daily between the hours of ten in the morning and three in the afternoon in Winter and ten in the morning and four in the afternoon in Summer Sundays and Holidays excepted. Third That the Vice Chancellor the two Proctors the Regius Professor of Medicine and the Keeper of the Ashmolean Museum of the said University and their successors in the same several offices for the time being and also the Reverend Richard Greswell of Worcester College B.D. and Henry Wentworth Acland of All Souls College in the said University Doctor of Medicine and their successors to be appointed from time to time as hereinafter mentioned shall be Curators And upon the death refusal or incapacity to act of the said Richard Greswell and Henry Wentworth Acland or either of them then a new Curator or two new Curators as the case may be shall be nominated by the surviving or other curators such nomination to be approved by the University in Convocation and so from time to time as often as any vacancy shall occur in any or either of the non-official Curators it being the intention that there shall always be two other Curators in addition to the five official Curators. Fourth That the Curators for the time being shall have full power and authority from time to time as they shall see fit to frame rules and regulations for the safe custody and preservation of all the property and for the management of and access to the said Collection and for the use of the Library but in the event of a managing Curator or Inspector being appointed for the better custody and preservation of the said Collection the consent of the said Frederick William Hope during his life time shall be necessary to such appointment. Fifth That the Curators shall also have power on obtaining the consent thereto of the said Frederick William Hope during his lifetime and after his decease of their own authority to dispose of any Duplicates in the Entomological Collection to any Museum Institution or person upon such terms and in such manner as they shall think proper But that this power shall not extend to the disposing of any other Specimens comprized in the said Collection and with a view to prevent any infested insect being added to and injuring the Collection no addition shall be made thereto unless previously approved of by the said Frederick William Hope or some competent person under the authority of the Curators. Sixth That if a new University Museum shall be established in Oxford and a Library formed in connection with it the Curators shall have power to place the whole of this Collection in suitable Rooms to be provided in such Museum but the Library and the Plates and Engravings shall in any case be under the entire control of the Curators who may dispose of the same in any way which they shall consider most conducive to the study of Natural History within the University as well as to the use of the same in connection with the Entomological Collection. Seventh That in all the more important proceedings relative to this Collection such as the framing new rules or regulations the appointment of Curators and the disposition of Duplicates or of the Library the concurrence of four at least of the Curators present at a Meeting to be called for any such purpose shall be requisite Provided always nevertheless and these Presents are upon this express condition that if the said Chancellor Masters and Scholars of the said University shall not by vote of Convocation within the space of twelve calendar months from the date hereof accept the said Collection articles and effects upon the terms and conditions hereinbefore set forth or if after such acceptance of the said Collection articles and effects and at any time during the life time of the said Frederick William Hope or the period of ten years from his decease there shall be any breach or failure in the performance of the said terms and conditions then and in either of such cases these presents and every thing herein contained shall cease determine and be absolutely void. And the said Collection articles and effects shall revert to and become the property of the said Frederick William Hope or his representatives. And the said Frederick William Hope doth hereby for himself his heirs executors and administrators covenant and agree with and to the said Chancellor Masters and Scholars and their successors that he the said Frederick William Hope his executors or administrators shall and will from time to time and at any time hereafter make do and execute any further gift grant bargain sale or other confirmation of these presents by or on behalf of the said Chancellor Masters and Scholars or their successors [as] shall or may be reasonably required In Witness whereof the said Frederick William Hope hath hereunto set his hand and seal this fourth day of August in the year of our Lord one thousand eight hundred and forty nine.

Appendix D. Transcript of Manuscript Oxford University Archives W.P. α/40/5

Report of the Committee on Mr. Hopes proposed Benefaction to the University.

The attention of the Committee has been directed
  1. To the Conditions of the Benefaction,
  2. To the Responsibilities incurred by its acceptance.

1. The Committee are of the opinion, that some of the Conditions should be modified; and, as the Collection must remain for a while in a temporary apartment, where some of them would be impracticable, they should be, for that period, partially suspended.

   They have accordingly instructed Mr. Morrell to prepare a kind of Supplemental Deed, with provisions for the above purposes, to which Mr. Hope might, (if the University be disposed to accept his benefaction) give full authority by his signature. The Draft will be submitted to the Board.

2. The Responsibilities of the University, incurred by the acceptance of this Benefaction, refer (1st) to the necessary accommodation for the present reception, and permanent lodging, of the Collection. (2^{d}) to the Expenses involved in its care and due Exhibition, and (3^{d}) to the continued Renewal of the Collection, as the present Specimens, which are many of them necessarily of a perishable nature, shall suffer by decay.

1. An apartment in the Taylor Institution has been readily afforded by the Curators for a limited period; but the Committee are at a loss to name any existing room, belonging to the University, suitable for the permanent lodgment of such a Collection. The University therefore by its acceptance will be bound to *erect* a suitable Edifice for the purpose.

2. The Care and Exhibition of the Collection will involve further liabilities, and these duties obviously cannot be satisfactorily discharged, without some considerable annual expense. There must be a person, above the lowest rank, to attend to, and exhibit, the Collection to Visitors; nor would it be creditable to the University to appoint a person to this office, with little or no knowledge of the subject, and his salary must of course be in proportion. It appears indeed that Mr. Hope has intimated his intention should he survive his Father, to contribute towards this expense.

3. Further, there is the responsibility of keeping up the Collection. For the University could not with any regard to its character, after accepting, suffer the Collection to waste or perish. What however may be the expense of purchasing new specimens in place of those, which decay, in order to preserve to the Collection its present reputation, or what Functionary of the University may be competent to give his attention to it and from time to time make the needful purchases, the Committee are not prepared to state.

Appendix E. Transcript of Manuscript Oxford University Archives W.P. α/40/4

The Board of Heads of Houses [& Proctors]† consider it their duty to communicate with Mr. Hope on the subject of the Conditions contained in his proposed deed of Gift to the University, in explanation of the sense in which the Members of the Board interpret them, and in which alone they could recommend to the University to accept the obligation.

In so doing they desire, in the first place, to express their high sense of the liberality & public spirit which, they feel assured, induces Mr. Hope to propose that his valuable Collection should be deposited in Oxford for the benefit of Students in his own favourite department of Natural History, and they hope that, in any objections they may make to the strict terms of either of the Conditions, or in any proposed qualification of them, no end or object will be imagined to exist on their part which is inconsistent with an honest & sincere intention to render this donation effectual for the avowed purpose of the donor.

As to the First Condition, the Board think it right to repeat, what they understand has been already explained to Mr. Hope personally by the Vice-Chancellor, that the University has no distinct Building at its disposal which can be allotted as an Entomological Museum: Also, they feel it right to say that, whatever possibility, or even probability, there may seem to be of other Buildings being erected, the University must not be held pledged, by its acceptance of this Collection, to any scheme of building. The utmost for which they can engage, and that only with the consent of the Curators of the Taylor Building, (which however has been obtained for the purpose) is that, as a temporary arrangement, the Collection shall remain in the room in which it is at present deposited, where it is safe, & where there is space for the disposition of the Cabinets & for the partial display of the specimens.

As to the Second Condition, the Board desire to say that, with the sincere intention on their part of making this Collection useful & accessible, free from the payment of any Fee upon any pretext whatever, they must be distinctly understood to reserve to the University [in Convocation] the absolute right of fixing what shall, from time to time, appear to be proper hours and regulations for the admission either of Strangers or of Members of the University.

As to the Third Condition, the Board are satisfied [have no desire to propose/suffer any alteration].

As to the Fourth, the Board approve that the Curators named in the deed, & their Successors, s^d have full & Entire control & management of this Collection: But it seems to them necessary to declare distinctly that their authority will not extend beyond the room in which the Collection is deposited, but must be subordinate to that of the Curators of the Building of which that room forms a portion, [as well as] also to that of the *Governing Body* [Convocation] of the University in respect of rules & regulations as to access.

The Appointment of a managing Curator, or Inspector, being alluded to in this Fourth Condition, (Such Inspector being of course understood to be a scientific person to whom an adequate Salary must be paid) the Board are not prepared at present to make any engagement on the part of the University: But it seems to them highly important to the safe custody & preservation of the property, as also to it's usefulness,

† *Note*: Items in square brackets represent contemporary annotations by Westwood and others.

that some such Inspector s.^d be appointed, and they would be glad to receive Mr. Hope's opinion on this point, & on the means of effecting it.

As to the Fifth Condition, the Board are satisfied. [have no alteration to suggest].

As to the Sixth Condition, they w.^d only observe that, in the *possible* event of this Collection being transferred to a new University Museum, the Authority of Mr. Hope's Curators, although absolute and distinct so far as regards their own property, will be held subordinate to that of the Curators of the Museum and liable to all the general rules necessary for the regulation of such an establishment.

As to the Seventh Condition, the Board are satisfied.

From the preceding observations it naturally results that the Board would wish to modify the Conditions proposed in the deed of Gift; and accordingly they beg to submit the following alterations for Mr. Hope's consideration Viz:

That the *First* Condition s.^d stand thus: That the said Collection, Library, Articles, & Effects, shall be deposited in a suitable Room, or Rooms, within the University, especial care being taken for their safe Custody & preservation, & for the suitable arrangement & display of the specimens, so far as [may prove practicable] the space will admit—

Second,
'That the said Collections Library Articles & Effects shall be made practically useful as a means of extending and improving a knowledge of the Entomological Department of Natural History and for that purpose the rooms shall be at all reasonable times free of access to Members of the University & other persons especially to learned Naturalists & other scientific persons of foreign Countries' subject however to such regulations as the Curators for the time being shall think fit, which regulations also *are* to be at all times under the control of the University [in Convocation].

Fourth,
'That the Curators for the time being shall have full power & authority from time to time as they shall see fit to frame rules & regulations for the safe custody and preservation of all the property and for the management of and access to the said Collection and for the use of the Library' so far as may not be inconsistent with the general rules & regulations established by the Curators of the Building of which the room or rooms in which such Collection & effects are [or may be] deposited [that may] form a part—

Sixth,
'That if a new University Museum shall be established in Oxford and a Library formed in connection with it the Curators shall have power to place the whole of this Collection in suitable rooms to be provided in such Museum, but the Library and the Plates and Engravings shall in any case be under the entire control of the Curators but subject to and so as in no case to conflict with any general rules & regulations which may at any time be established [by the University in Convocation] for the management of such Museum, and the said Curators shall have power to dispose of the same Library Plates & Engravings in any way which they shall consider most conducive to the study of Natural History within the University as well as to the use of the same in connection with the Entomological Collection'.

In Conclusion, the Members of the Board assure Mr. Hope that the alterations which they have thus suggested do not originate, on their part, in any [other] feeling (of jealousy) (but in) [than that of] a sincere desire that this valuable Collection, when formally accepted, may become practically serviceable: And they take the liberty of [expressing their conviction] reminding Mr. Hope that, according to all experience, such a desirable event will be best promoted, not so much by strictness of legal conditions, which may possibly be evaded, as by a liberal expression of the spirit & intention in which a gift of this nature is presented & accepted.

Index of names

(Note: this index covers the main text only (pp. 1–65))

Acland, H. W. 6, 11, 12, 15, 16, 18, 23, 24, 25
Audouin, J. V. 42–3, 63

Baldwin, J. M. 48
Bell, T. 18
Bingham, C. T. 59
Boheman, C. H. 43
Boulenger, G. A. 12
Brongniart, A. 26
Brullé, G. A. 43
Burstal, E. 44

Carpenter, G. D. Hale 28, 31–2, 49–52
Charpentier, T. de 42
Chattaway, D. F. 28
Chiaje, S. Delle 23–4
Cocco, A. 24
Collins, J. J. 30, 58–9
Cross, L. B. 48
Curtis, J. 37
Cuvier, G. C. 42

Dallas, W. S. 25
Darwin, C. 2, 18, 38, 46
Daubeny, C. 17–18
De Saram, S. H. 2
Dillenius, J. J. 12
Disraeli, B. 2
Dixey, F. A. 56
Douglas, J. W. 29

Edwards, H. 39
Eltringham, H. 47, 59
Evans, W. F. 3

Fabricius, J. C. 43
Faraday, M. 18
Fortnum, C. D. E. 2, 11, 25
Fortune, R. 2

Geldart, W. M. 63
Godman, F. D. 59
Gradwell, G. R. 52
Gray, G. R. 37, 42
Gray, J. E. 18
Grensted, L. W. 61
Greswell, R. 5, 7, 8, 16–17, 19
Guérin-Méneville, F. E. 42
Guilding, L. 1–2

Haliday, A. H. 20
Hamm, A. H. 30–1, 56–7
Hanitsch, R. 60, 64
Hassell, M. P. 52
Haworth, A. H. 36–7, 63
Hazel, A. E. W. 48
Herrich-Schäffer, G. A. W. 43
Heyden, C. H. G. 42
Higgins, E. 16, 19
Higgins, L. G. 64
Hobby, B. M. 61
Holland, W. 30–1, 56, 59
Hope, Mrs Ellen (née Meredith) 2, 9, 21–2, 25, 46
Hope, F. W. 1–9, 11–12, 15–21, 23–7, 33, 36–7, 39, 42, 63

Jackson, W. Hatchett 45
Janson, O. E. 27
Jenyns, L. 18
Jeune, F. 17
Jones, J. Viriamu 48
Jussieu family 26

Kidd, J. 1, 5, 8

Lack, D. 51
Lamborn, W. A. S. 31
Lankester, E. R. 48
Latreille, P. A. 42
Lee, J. 11

 Index of names